MARINE POLLUTION

WHAT EVERYONE NEEDS TO KNOW®

MARINE POLLUTION

WHAT EVERYONE NEEDS TO KNOW®

JUDITH S. WEIS

OXFORD
UNIVERSITY PRESS

OXFORD

UNIVERSITY PRESS

Oxford University Press is a department of the University of Oxford.
It furthers the University's objective of excellence in research, scholarship,
and education by publishing worldwide.

Oxford   New York

Auckland   Cape Town   Dar es Salaam   Hong Kong   Karachi
Kuala Lumpur   Madrid   Melbourne   Mexico City   Nairobi
New Delhi   Shanghai   Taipei   Toronto

With offices in

Argentina   Austria   Brazil   Chile   Czech Republic   France   Greece
Guatemala   Hungary   Italy   Japan   Poland   Portugal   Singapore
South Korea   Switzerland   Thailand   Turkey   Ukraine   Vietnam

Oxford is a registered trademark of Oxford University Press
in the UK and certain other countries.

"What Everyone Needs to Know" is a registered trademark of
Oxford University Press.

Published in the United States of America by
Oxford University Press
198 Madison Avenue, New York, NY 10016

Library of Congress Cataloging-in-Publication Data
Weis, Judith S., 1941–
Marine pollution : what everyone needs to know / Judith S. Weis.
pages cm.
Includes bibliographical references and index.
ISBN 978–0–19–999669–8 (hardcover : alk. paper) —
ISBN 978–0–19–999668–1 (pbk. : alk. paper)
1. Marine pollution.   2. Marine ecology.   I. Title.
GC1085.W45 2015
577.7'27—dc23
2014009013

1 3 5 7 9 8 6 4 2
Printed in the United States of America
on acid-free paper

CONTENTS

3. Marine Debris   42

4. Oil and Related Chemicals   63

5. Metals 83

8. Bioaccumulation and Biomagnification   143

9. Climate Change and Ocean Acidification   163

PREFACE

Many people throughout the world were horrified to read daily reports about the huge volume of oil spewing from the drilling rig Deepwater Horizon in the Gulf of Mexico for many months in 2010. People were similarly riveted reading the news and seeing on TV photos of the oil-covered birds and sea otters in Prince William Sound, Alaska after the *Exxon Valdez* spill in 1989. These spectacular tragic events are fortunately rare. People may also become aware of marine pollution through some smaller events such as a fish kill in a local area, excessive debris or seaweed littering the beach, or discolored water from an algal bloom. These visible signs of marine pollution are not all there is. There are many types of pollution that have no visible signs and are only detected by sophisticated chemical analysis. In this case, what you don't know can sometimes hurt you—and if it doesn't hurt us, it might hurt marine organisms.

The marine environment is under assault from overfishing, habitat loss, and pollution. New kinds of pollutants ("contaminants of emerging concern") include both new pollutants and old pollutants that no one ever paid attention to before. These include pharmaceuticals that are designed to have effects on the body at very low concentrations. The unsightly volumes of marine debris, mostly plastic, washing up on beaches and collecting in great garbage patches in the oceans is something that most people have heard about. Marine debris made the headlines in March and April 2014 when the search and rescue teams seeking the missing Malaysia Airlines Flight 370 discovered that the ocean is full of garbage. When ships were able to retrieve "suspicious" items that planes had spotted, it turned out not to be debris from the missing plane, but ordinary garbage swirling around in the ocean.

New awareness of the damaging effects of loud noise on marine animals, especially mammals, is of great concern, as it may relate to whale beaching incidents. There has been increasing concern and attention in recent years to the effects of ocean acidification, caused by increased levels of carbon dioxide in the atmosphere. While much of the concern and attention is about impaired shell formation, people are also discovering effects of ocean acidification on physiology and behavior as well. Perhaps the most widespread and serious type of pollution worldwide is eutrophication due to excess nutrients, which stimulate algal blooms and reduce the level of oxygen. While eutrophic areas and "dead zones" are increasing around the world, there is also some good news in that many persistent organic pollutants have been banned and are no longer manufactured (even though they still remain in sediments and accumulate in marine life). Also, the frequency of oil spills has gone down in the past few decades. In addition to this reduction of inputs of some historical pollutants, efforts have begun to physically remove highly contaminated sediments from some of the estuarine toxic hot-spots in the United States under the auspices of the Superfund Program.

This book, like others in the What Everyone Needs to Know® series, is intended for the general public, including policymakers, naturalists, environmentalists, students, and scientists in other fields. I hope it will provide greater understanding and stimulate greater interest in the topic, and I hope that a more educated public will strongly support taking action to reduce marine pollution. In this book I cover the visible and the invisible types of marine pollution—where it comes from, what it does, and how we might be able to reduce it. Chapters are organized by type of pollution. In addition to the usual types of pollution, there is a chapter dealing with invasive species, not always considered a type of pollution, under the category of biological pollution. I also have a chapter about climate change—comprising global warming, sea level rise, and ocean acidification—and effects on marine life. Within each chapter I include questions that you may have thought about, including potential effects of the pollutants on our own health, and many questions you may not have wondered about, including topics such as the fate of chemical pollutants in the marine environment, what effects pollutants have on marine organisms, and how marine organisms cope with different types of pollutants. I hope that in both cases you will find the answers interesting and useful. Perhaps they will stimulate you to think of additional questions that you would like to know about. The final chapter covers prospects for the future and includes sections on international and national laws regulating pollution, how states and municipalities can reduce pollution, and steps that individuals can take to reduce pollution. A large number of suggestions are provided on how you can make a difference in reducing marine pollution.

ACKNOWLEDGMENTS

I would like to thank my husband, Dr. Peddrick Weis, for his valuable suggestions as I was writing this book, his assistance with the figures, and his role as a frequent research partner during many years of studying effects of pollution on marine organisms. I am very grateful to Rachel Carson for her books about the sea that fostered my interest in marine biology, and for writing *Silent Spring*, which stimulated my interest in pollution. I also thank John and Winona Vernberg, Anthony Calabrese, and Fred Thurberg, who organized a series of conferences on marine pollution in the 1970s and 1980s that were instrumental in guiding my early research directions in the field. The many graduate students and postdocs who worked in my lab on pollution-related research topics have contributed a great deal. I thank Jeremy Lewis of Oxford University Press for his encouragement and sound advice throughout the process of creating this book. I am also grateful to the governmental and nongovernmental environmental organizations that are working to reduce pollution in the oceans and elsewhere.

MARINE POLLUTION

WHAT EVERYONE NEEDS TO KNOW®

1

INTRODUCTION TO THE MARINE ENVIRONMENT AND POLLUTION

What is the marine environment?

As used in this book, the marine environment covers not only the ocean, but estuaries (e.g., bays), which are coastal areas where the seawater is diluted with freshwater coming from rivers and streams, or sometimes groundwater. Much of the pollution is concentrated in these shallow coastal areas, which are often next to urban centers and other concentrations of humans who are responsible for the pollution.

What are some basics of marine ecosystems and food webs?

Marine ecology is a branch of ecology dealing with the interrelations of organisms living in the oceans, shallow coastal waters, and on the sea shore. Organisms interact through the roles they play as producers, consumers, and decomposers. Primary producers are plants that take in inorganic carbon dioxide and water, and through the process of photosynthesis make organic materials (sugars) using light energy from the sun. They are the first step of the food web. Primary consumers are herbivorous animals that eat the plants; secondary consumers are carnivorous animals that eat the herbivores;

third-level consumers are carnivores that eat other carnivores; and decomposers are microorganisms (such as bacteria and fungi) that break down the organic materials from the plants and animals (excretory products and dead bodies) into inorganic materials, which are eventually reused by producers. The decomposers are concentrated in the sand or mud on the bottom, and play an essential role in recycling materials. There are more producers than consumers, more primary consumers than secondary consumers, and so on up the chain, because at each step in the food chain a great deal of energy is lost—it is not efficient. So top carnivores (for example sharks) are the rarest animals.

The most important primary producers in the ocean are a diverse group of microscopic floating single-celled photosynthetic organisms called phytoplankton. They are the basis of the food web that supports the rest of oceanic life. They are widely distributed in huge numbers, but occur near the surface of the water only down as far as light penetrates, since light is essential for photosynthesis. Phytoplankton are eaten by small floating animals called zooplankton. Zooplankton consist of a wide variety of different types of generally small animals, some of which spend their whole life as small plankton, while others are larval stages of larger animals such as clams or crabs that will subsequently go to the bottom to live as adults. Zooplankton, in turn, are eaten by small fish, which are eaten by larger fish, which may be eaten by very large fish (or other large animals such as marine mammals). Animals that live on the bottom are called benthos; some benthic animals obtain their food by filtering the plankton, while others consume decaying plant or animal material (called detritus) that sinks down to the bottom.

In shallow coastal areas or estuaries, additional kinds of primary producers are found: larger algae (seaweeds) or rooted plants like seagrasses that live attached on the bottom, since the light can penetrate through the shallow water. These

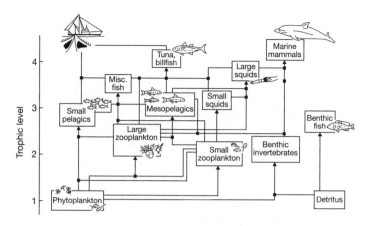

Figure 1.1 Marine food web showing different trophic levels (from Wikimedia)

are consumed by various animals, but mostly after they have died and decayed into detritus (Figure 1.1).

Why is there concern about the state of the oceans?

For centuries, it was thought that the oceans were so vast that nothing people could do could possibly have an impact on them. However, contrary to this belief, it turns out that we have been doing so for many years. Back in 1951, Rachel Carson wrote in *The Sea Around Us* that people could not change the ocean the way we have plundered the continents, but she subsequently changed her opinion. We are now aware that many fish populations are declining from over-fishing, that warming is melting the polar ice and raising sea levels, and that portions of the ocean are full of trash—plastic bottles and bags, balloons, and lost fishing nets. We have witnessed disastrous oil spills. We find abnormalities in marine animals due to subtle effects of man-made chemi-cals and find large coastal areas with water devoid of oxygen, and therefore of marine life, due to wastes released into the waters.

Marine ecosystems are very important for the health of both marine and terrestrial environments. Coastal habitats account for about one-third of all marine biological productivity, and some estuarine ecosystems (i.e., salt marshes, seagrasses, mangrove forests) are among the most productive regions on the planet. In addition, other marine ecosystems, namely coral reefs, provide food and shelter to the greatest amount of marine biological diversity in the world. The ocean plays a key role in cycles of carbon, nitrogen, phosphorus, and other important chemicals. Ocean chemistry has been changing due to human activities both in coastal waters and in the open ocean. Some of the greatest impacts are on carbon, nitrogen, and dissolved oxygen, which affect biological functioning. Decades of pollution, along with destruction of coastal habitats and overfishing, have had devastating impacts on marine biodiversity and habitats. The increasing demand for seafood worldwide has depleted many fish populations, along with the economies of some coastal communities. On top of this, climate change is altering the oceans in ways that we are only beginning to understand. There is growing scientific evidence demonstrating serious—sometimes disastrous—impacts of pollution in the marine environment. Chemical pollutants of greatest concern are those that are widespread and persistent in the environment, accumulate in organisms, and cause effects at low concentrations. Toxic chemicals are varied and often difficult to detect.

What is a contaminant? Is there a difference between a contaminant and a pollutant?

A contaminant is a biological, chemical, or physical substance or energy normally absent or rare in the environment, which is present and which, in sufficient concentration, can adversely affect living organisms. A pollutant is substance or energy introduced into the environment that has undesired effects. So if a contaminant is present in high enough concentration, it is a pollutant. It could be something that occurs naturally in the environment (e.g., metals) but is in excess, or could be something that is man-made. Pollutants may be classified by their origin, by their effects on organisms, by their properties (such as toxicity), or by their persistence in the environment. Toxic chemicals are very varied, numerous, and expensive to monitor.

What are the major sources of pollution in the marine environment?

Land-based sources pollute estuaries and coastal waters with nutrients, sediments, and pathogens (disease organisms), as well as potentially toxic chemicals including metals, pesticides, industrial products, and pharmaceuticals. Following the Industrial Revolution, more and more material has been discharged from industries, sewage treatment plants, and agriculture, eventually reaching marine ecosystems. But pollution does not come exclusively from land-based sources. Highly visible events such as the *Exxon Valdez* oil spill in Alaska and the Deepwater Horizon gusher in the Gulf of Mexico have polluted the seas with oil from ships, and from drilling platforms in the ocean itself. These highly publicized events have raised public awareness of marine pollution. Other water-based sources of pollution are less spectacular, and include discharge of waste from vessels, the leaching of antifouling paints from ships, and leaching of wood preservatives (e.g., creosote or chromated copper arsenate) from wooden bulkheads and dock pilings. Aquaculture operations such as floating cages in which salmon are raised can pollute nearby waters with fish wastes, uneaten food, antiparasite chemicals, and antibiotics. Pollution can also enter the ocean from the atmosphere. For example, the metal mercury is released as a gas into the atmosphere from burning coal, and subsequently can be deposited in the oceans. Nitrogen, in the form of nitrogen oxides from the burning of fossil fuels, is also an air pollutant before being deposited into the ocean in precipitation and becoming a water pollutant.

What are the major ways that land-based pollutants enter the marine environment?

"Ocean dumping" refers to transporting materials in a barge and physically dumping them in the ocean. The dumping of industrial, nuclear, sewage, and many other types of waste into oceans was legal in the United States until the early 1970s, when it became regulated; however, dumping still occurs illegally everywhere. The movement to ban ocean dumping of sewage sludge gained momentum in the United States when contaminated wastes from sewage-derived microorganisms were discovered at public beaches, along with unsavory items such as hypodermic syringes and tampon applicators. Most of the chemical pollution in the ocean comes into the water through pipes rather than dumping. While many pollutants are discharged (legally) from industrial or residential areas, others come from agricultural areas. Factories and sewage treatment plants release their wastes into receiving waters through a pipe, referred to as a "point source," which can be monitored and regulated by environmental protection agencies. Since passage of the Clean Water Act in the United States in 1972, much progress has been made in controlling pollution from point sources. Combined sewer overflow (CSO) occurs in older cities, however, where storm drains connect to pipes going to sewage treatment plants from homes and industries. Heavy rainfall can overwhelm the capacity of the sewage treatment plants, causing everything to go out into the water untreated. The resulting bacterial contamination from sewage leads to beach closures for health reasons.

In recent decades attention has moved from point sources to diffuse runoff and atmospheric deposition (called "nonpoint sources"). Contaminants that wash into the water from soil,

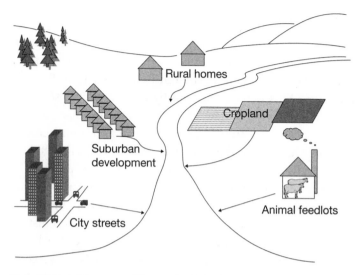

Figure 1.2 Nonpoint source runoff from rural and urban landscapes (permission from Dr. Peddrick Weis)

streets, construction sites, and so on during rainfall can enter water bodies in many places, as do pollutants from the atmosphere that come down in rainfall. This pollution is not so easily regulated. Nonpoint sources such as farms, roadways, and urban or suburban landscapes remain largely uncontrolled, and are major sources of continuing pollution inputs (Figure 1.2). If it is not directed to sewage treatment plants causing CSO, urban stormwater runs directly into water bodies, bringing with it sediments, grease, litter, oil, polycyclic aromatic hydrocarbons (PAHs), and metals from highways.

Which pollutants enter the ocean from the air?

Nitrogen gases, mercury, carbon dioxide, and radioactive isotopes come largely from the atmosphere. Some organic chemical pollutants (e.g., polychlorinated biphenyls, or PCBs) can also be transported long distances in the air before being deposited in the ocean.

Can objects in the water cause pollution?

Antifouling paints on vessels are designed to reduce attach-
ment of organisms like barnacles and algae, and do so by
being toxic. The chemicals are released slowly from the boat
paint and thus deter settlement by the planktonic stages of
these organisms. However, the chemicals are also toxic to
other organisms nearby. The most popular antifouling chemical that was used in the past was tributyltin (TBT), which is
now banned throughout most of the world (and will be discussed in detail in later chapters). Other antifoulants include
copper, which is especially toxic to mollusks and algae (it is
used as an algicide and molluscicide). Since bans and restrictions on TBT came into effect, researchers have developed
and produced new types of chemicals. Irgarol is now a common antifoulant, and is highly toxic to nontarget plants. It is
found in water and sediments near marinas at levels that may
be high enough to cause changes in phytoplankton communities. Another antifouling biocide, diuron, is also found in
water and sediments.

When wooden structures are placed in the water in the
form of dock pilings or bulkheads, they are subject to decay
by microbes and destruction by wood-boring animals such
as some amphipods (gribbles) and shipworms (which are
really mollusks). Therefore, the wood gets treated with
high concentrations of toxic chemicals, such as creosote or
chromated-copper-arsenate (CCA), to prevent this destruction. These chemicals also leach from the wood and can accumulate in the environment and get taken up by nearby plants
and animals, causing toxic effects.

How can aquaculture cause pollution?

Aquaculture is the raising of marine organisms for food—
farming the sea—similar to agriculture on land. Fish farms,
especially open cage culture of salmon, have been found to

be sources of pollution in local waters. Thousands of fish concentrated in open net pens produce tons of feces. Combined with uneaten food, this waste sinks to the bottom and affects the local environment, polluting the water and smothering plants and animals on the seafloor below the cages. For example, the nutrients in unused fish feed and fish feces can cause local algal blooms, which lead to reduced oxygen in the water, which in turn can lead to the production of ammonia, methane, and hydrogen sulfide, which are toxic to many aquatic species. Low oxygen can also directly kill marine life. Many types of aquaculture use chemical treatments such as antibiotics or antiparasite chemicals for a successful harvest. The amount of these chemicals released into the environment determines their effects on other organisms. A wide range of chemicals is currently used in the aquaculture industry—primarily pharmaceuticals such as antibiotics and antiparasitic chemicals, and antifouling agents such as copper for the cages. In some areas, such as Southeast Asia and South America, overuse of antibiotics has led to increased resistance of bacteria to treatment, which can make them much more harmful to the cultured species and potentially to other species, including humans.

Once in the water, what happens to the pollutants?

Ocean currents and organisms may redistribute pollutants considerable distances. However, sediments tend to bind metals, and many organic contaminants concentrate in the bottom sediments. The historic use of some chemicals that are no longer manufactured in the United States (e.g., DDT, PCBs) has left a legacy of contamination in the sediments, which remain contaminated with these persistent chemicals that continue to cycle through the environment and affect marine life decades after their input has ceased. Contaminated sediments also pose a problem for dredging operations, because the dredging process can release the contaminants from the sediments and make them more available to biota. Another thorny issue is where to put the contaminated sediments once they have been dredged up from the bottom. Solving these problems is a major reason for long delays in dredging for deepening ship channels and for cleanups of toxic hot spots. Organisms can take up or bioaccumulate chemicals from the environment. Once taken up into the body, the chemicals can exert toxic effects.

How do chemicals get into marine animals?

Aquatic animals take pollutants into their body through the skin, gills, and digestive tract, and excrete them in their waste or expel them through the gills. When the rate of uptake is greater than removal, the chemical builds up in the body. Chemicals that have low solubility in water and bind to sediments tend to accumulate to greater concentrations in organisms, especially in their fatty tissues. Chlorinated hydrocarbon pesticides, polychlorinated biphenyls (PCBs), and methylmercury are among those toxic substances with low water solubility that concentrate in organisms and are not readily metabolized or excreted.

Contaminants are transferred through food webs from prey to predator (trophic transfer), and some chemicals tend to become more concentrated during this process—a phenomenon called biomagnification. Persistent organic chemicals like PCBs and DDT, as well as methylmercury, tend to build up or biomagnify as they go from prey to predator, causing the largest, long-lived top predator to have the highest levels (Figure 1.3). An animal in a polluted area accumulates toxic chemicals from each item of contaminated food that it eats; concentrations are higher in consumers than in their food, and are highest in the top carnivores such as large fish, fish-eating birds, marine mammals, and humans. Because of biomagnification, methylmercury levels can be quite high in large carnivorous fish like swordfish and tuna, even though they live in the open

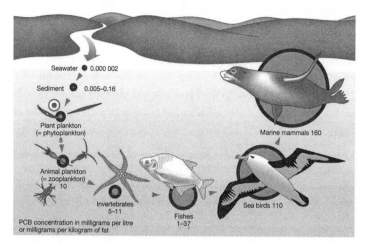

Figure 1.3 Biomagnification of contaminants up the food web (© Walther-Maria Scheid, Berlin, Germany, for World Ocean Review 2010)

ocean far from any source of mercury. It is recommended that people (particularly pregnant women and young children) not eat a lot of these fish. Chlorinated pesticides, PCBs, and dioxin also undergo biomagnification, but metals other than methylmercury do not do so.

The sex of a fish may affect how much of a contaminant it accumulates. Egg yolk is a fat-rich substance that can store large quantities of organic contaminants, and some females put large amounts of these fat-soluble chemicals into eggs, reducing the levels in their bodies. This maternal transfer of contaminants is found in egg-laying birds and reptiles as well as fishes. While it is a good for the females to reduce their own pollutant level, it certainly does not benefit the offspring to start out their lives already loaded with toxic chemicals.

What is toxicity?

A toxic substance is one that harms living things at low concentrations. (Almost anything can be harmful if there is enough of it!) Toxic effects have been studied primarily in laboratory experiments (bioassays), although there have been some field studies of effects on populations of marine organisms. Early studies of pollutant effects relied on tests that measured lethality (death). The LC_{50}—the concentration of a chemical that caused 50% of the test animals to die (typically in 96 hours)—was the benchmark. Regulations under laws such as the US Federal Insecticide Fungicide and Rodenticide Act (FIFRA) for developing safe levels of pesticides to protect aquatic life require the standard LC_{50}, which is of little ecological relevance. Toxicity tests are required for a few species: rainbow trout, bluegill, and daphnids—one cold-water fish, one warm-water fish, and one crustacean, all freshwater species. Unfortunately, even today, over half a century later, this approach—measuring what percentage of the animals die in 96 hours—is still considered most useful in a regulatory context. These tests do not consider sublethal effects that occur over longer periods of time, or toxicity that is delayed, or differences in life history among species. Knowing about effects of longer-term, lower-dose exposures on physiology, behavior, and development is essential for understanding overall impacts of pollutants in nature.

What effects can pollutants have besides killing living things?

Extensive research has shown that toxic chemicals can disrupt metabolic, regulatory, or disease defense systems, and reduce reproduction. Behavior, development, and physiology are all sensitive to pollutants. Learning about these sublethal effects can help us understand the mechanisms of action of different chemicals, and also to understand ecological effects in the real world. We have learned that early life processes and stages— eggs and sperm, fertilization, embryonic development, and larvae—are very sensitive to contaminants, so setting "safe" levels based on how much of a chemical will kill adults will not protect these young stages. The hormonal control of reproduction can be affected by many chemicals, now called "endocrine disruptors." Exposures during early life may cause effects that do not appear until later, sometimes many years later. Thus, long-term delayed effects and indirect effects are important. There has been some progress toward greater ecological realism, but advances have been mainly in freshwater ecosystems.

The effects of chemicals on individuals may cause changes in populations and result in reduced population growth rate, lower population size, reduced birth rates, and higher death rates, producing a population dominated by younger, smaller individuals with reduced genetic variability. Reduced genetic variability happens when the more susceptible individuals disappear from the population and the more pollution-tolerant ones become predominant, as has been seen with insecticide-resistant insects or antibiotic-resistant bacteria.

Toxic effects appear first at the biochemical level, and later at the cellular level, then the level of the whole organism, the population, and eventually the ecological community as a whole. Initial biochemical changes observed can be altered enzymes, changes in DNA and RNA, or the production of particular proteins that can detoxify the chemical. At the cellular level, chromosome damage, cell death, abnormal structures, or cancer can occur. Some chemicals affect the immune system and increase susceptibility to infectious diseases. At the level of the whole organism, changes in physiology, development, growth, behavior, or reproductive capacity may occur, and at high concentrations, the animal or plant can die.

Fortunately, we have seen in many locations that when the input of pollutants decreases or toxic waste sites are cleaned up, the incidence of diseases and other problems diminishes. Tolerance to the contaminants may be lost as well. In a contaminated marsh near a former battery plant close to the Hudson River that released cadmium for decades, Jeffrey Levinton and colleagues from Stony Brook University showed that the worms in the sediments had become highly tolerant to cadmium. Some years after the pollution was cleaned up, as required by the Environmental Protection Agency (EPA), the scientists revisited the site and found that the worms had lost their cadmium tolerance over relatively few generations.

How is the degree of toxicity measured?

"The dose makes the poison." It is important to have accurate measurements of how much of a given chemical causes a given effect. Contaminants generally occur in low concentrations, but small concentrations such as parts per million and parts per billion can cause effects. A part per million (ppm) seems like a very small amount—and it is. One ppm (or mg/l) is equivalent to one drop of a substance in about 13.2 gallons of water. One ppb (or μg/l) is one part in 1 billion—much smaller than a ppm. One drop in one of the largest tanker trucks used to haul gasoline would be 1 ppb. Some chemicals, including dioxin and tributyltin are toxic at levels below 1 ppb. It is difficult and expensive to measure these low concentrations of contaminants. Sophisticated equipment such as atomic absorption spectrophotometers or gas chromatograph/mass spectrometers is needed.

How can field studies be used to understand toxicity?

Integrated field approaches are important, along with laboratory studies to provide insights into effects at the population and community level. Field experiments can investigate contaminated environments—but hardly ever, only under very restricted conditions, may scientists release known amounts of chemicals in the field to observe effects in controlled experiments. Attempts to bring the field closer to the lab include studies on multiple species placed together in microcosms (small containers) or mesocosms (large containers), which can be used to investigate community level effects of contaminants. They allow for replication, so dose-response relationships under controlled conditions can be studied. These kinds of studies can show the differential sensitivity of different species and can be used to learn about biological interactions. There is much to be learned from such approaches. However, dosing of complex mesocosms with known concentrations of specific chemicals still does not really duplicate the natural environment in which organisms are subjected to many different pollutants at different concentrations (which vary over time), and where some of the species may have evolved increased resistance to some contaminants. Thus, there remains uncertainty with ecological risk assessments and with translating mesocosm results to real-world field situations.

It is usually very difficult to attribute problems seen in the field to particular contaminants, because generally there are many different contaminants at a site. In some rare cases, observations on natural populations in the field called attention to effects of certain chemicals. This was the case with tributyltin's (TBT) effects on oysters in Europe (see Chapter 8). Since the abnormalities produced by TBT are unique and not produced by other chemicals, the causal connection between observed effects (abnormal shells in oysters) and the particular chemical (TBT) could be seen more easily.

Why are some species more sensitive to pollution than others?

Differences in sensitivity are due to differences in physiology, generation time, and life cycle among species, which can all affect initial responses and the ability to recover from the effects. Species that are short-lived and produce large numbers of offspring can exploit changing environments, including contaminated ones. Such species with short generation times also are more likely to be able to evolve tolerance to contaminants. High metabolic rates can lead to more rapid breakdown of pollutants. In contrast, species that are long-lived, slow to mature, and have relatively few offspring are less likely to be able to evolve resistance to contaminants. Also, long-lived species tend to be higher up on the food web, fewer in number, and to accumulate higher levels of contaminants over a long period of time. Their slow reproduction makes potential population recovery from declines very slow. Slow reproduction, combined with high accumulation of contaminants, makes them particularly vulnerable to reproductive effects. Transfer of fat-soluble contaminants (e.g., PCBs, DDT) from females into the yolk of developing eggs exposes the next generation to these chemicals even before they are hatched.

What laws regulate marine pollution?

The ocean, as well as marine pollution from land-based sources, is governed by legal frameworks at the international, national, state, and local levels. Multilateral and bilateral treaties and other agreements are in place for fishery management, shipping, protecting biodiversity, and pollution. The multinational treaty on pollution is the International Convention for the Prevention of Pollution from Ships, commonly known as MARPOL, which regulates discharges into the ocean. MARPOL is a comprehensive treaty that regulates pollution from ships. Six annexes to the treaty set out regulations for different aspects of pollution. Annex I covers prevention of pollution by oil from operational measures and from accidental discharges; Annex II regulates pollution by noxious liquid substances carried in bulk (some 250 substances were evaluated and included in the list); Annex III specifies requirements for the issuing of detailed standards on packing, marking, labeling, documentation, stowage, and quantity limitations for "harmful substances"; Annex IV contains requirements to control pollution by sewage (the discharge of sewage is prohibited, except when the ship has an approved sewage treatment plant or is discharging disinfected sewage using an approved system); Annex V governs garbage and bans discharge of plastic from ships; and Annex VI limits sulfur oxide and nitrogen oxide emissions from ship exhausts and prohibits emissions of ozone depleting substances into the air. MARPOL, administered by the International Maritime Organization (IMO), creates obligations for both flag states (the country certifying a vessel, from which a vessel launched, or under which a vessel sails) and port states (where a vessel lands). Both flag states and port states may inspect vessels to make sure they are in compliance with the treaty and can impose sanctions if it is in violation of the terms. In the United States, the Coast Guard has the primary responsibility.

Like marine-based sources, land-based sources are regulated by all levels of government. The United Nations Convention on the Law of the Sea (UNCLOS) is an international treaty that covers many aspects of ocean governance and includes obligations to control land-based sources of pollution. In addition to UNCLOS, regional treaties and domestic laws attempt to control land-based pollution. For example, the Cartagena Convention's Protocol Concerning Pollution from Land-Based Sources and Activities seeks to prevent land-based solid waste from coming into the Caribbean Sea. The terms of this treaty include preventing "persistent synthetic and other materials" from harming the ocean. Treaties like this provide both a legally enforceable framework and a forum in which countries can come together to exchange best practices and voluntary approaches to combat pollution.

In the United States, the Clean Water Act (CWA) seeks to control land-based sources of pollution. The CWA made it unlawful to discharge any pollutant from a point source (pipe or man-made ditch) into navigable waters unless a permit was obtained. It is enforced by the EPA. The EPA's National Pollutant Discharge Elimination System (NPDES) is a permit program that controls point source discharges into the aquatic environment. Individual homes that are connected to a municipal system, use a septic system, or do not have a surface discharge do not need an NPDES permit; however, industrial, municipal, and other facilities must obtain permits if their discharges go directly to surface waters. The CWA also provided funding for municipalities to construct or upgrade sewage treatment plants. The EPA has implemented pollution control programs such as setting wastewater standards for industry, and has set water quality standards for a large number of contaminants in surface waters. Beyond this, there are additional controls for waters that are impaired by pollution. Section 303 of the Clean Water Act authorizes states to identify impaired waters and calculate limits on the levels of various pollutants that can enter the impaired water. These limits are called total maximum daily loads (TMDLs). In 2007, California created a TMDL for the Los Angeles River in an attempt to reduce the amount of garbage entering that river, which would in turn reduce the amount of garbage entering the Pacific Ocean. The CWA will be discussed further in Chapter 11.

Why are some contaminants that have been banned still a problem?

National and international laws can regulate or ban chemicals, but "legacy pollution" from persistent contaminants (e.g., DDT, PCBs, metals) can remain in sediments for decades after their use or discharge has been banned, and sediments are a continuing source of contaminants to organisms. In addition, many pollutants are still not regulated, and there are inadequate controls on nonpoint sources. Environmental regulations and the level of compliance vary widely among countries. Nevertheless, much has improved in US waters as a result of the Clean Water Act, which stimulated many municipalities to build or upgrade sewage treatment plants.

How extensive and severe is marine pollution around the world?

While humans depend on the oceans for a variety of goods and services, we have altered and impaired the oceans both directly and indirectly. *A Global Map of Human Impact on Marine*

Ecosystems, a high-resolution map and atlas combining numerous data sets of the world's oceans, was published a few years ago by a large study group. It reveals that human activities have strongly affected approximately 40% of the marine area and have left only about 4% relatively pristine. It covers 17 different types of human activities, including climate change and fishing, as well as pollution. The authors compiled data from a variety of sources and fed them into a model that assigned a single number to each square kilometer of ocean, reflecting the overall human impact at that spot. The most highly affected marine areas are the eastern Caribbean, the North Sea, and Japanese waters, and the least affected ones are around the poles. The most heavily affected types of environments are continental shelves, rocky reefs, coral reefs, seagrass beds, and seamounts. There are few areas of coral reefs, mangroves, or seagrass beds in the world that are relatively unaffected. While not all affected areas are affected by pollution, many of them are. The major types of pollution are excess nutrients (eutrophication), marine debris, oil spills, and toxic contaminants.

2

NUTRIENTS

Why are nutrients considered pollutants, since they are required for life?

Input of excess nutrients such as nitrogen (N) and phosphorus (P) causes major problems in the aquatic environment. While phosphorus tends to be the main cause in freshwater, nitrogen is the major source of problems in the marine environment.

Where do the nutrients come from?

Sources of nutrients include sewage and food wastes plus animal wastes and fertilizers that are discharged or run off from agricultural areas. From land, excess N flows from agricultural fields, suburban lawns, and stockyards, entering freshwater and going down to estuaries via streams and rivers, altering water chemistry and ecology. As stormwater runoff flows over the land or impervious surfaces such as paved streets, parking lots, and building rooftops, it accumulates debris, chemicals, sediment, and other pollutants that can impair water quality. Urban areas contribute food wastes, human sewage, animal wastes, and lawn fertilizers. Even after treatment, sewage contains high levels of nutrients. Waste from septic tanks enters estuaries through seepage into groundwater. Wherever there is more residential development and more septic tanks in the neighborhood, more nitrogen seeps into nearby bodies of water, marshes, and estuaries. Nutrient enrichment due to excessive amounts of N is the primary cause of impaired coastal waters worldwide. Nitrogen is an essential nutrient and a fertilizer that is important for agricultural productivity, but when too much of it gets into the water it is a pollutant. The ability to synthesize N into fertilizer on an industrial scale increased crop yields throughout the twentieth century. Synthetic fertilizer not only fueled this growth, it also supported human population growth, providing a steady and cheap supply of grains. Synthetic fertilizer was a benefit in terms of crop yield but is an ongoing environmental problem, primarily because of nutrient runoff into aquatic ecosystems. The increased use of commercial fertilizers has increased N inputs by tenfold in many parts of the world. Only about 18% of the N in fertilizer actually gets into the produce; the rest is absorbed in the soil, runs off into the water, or enters the atmosphere. The amount of manure produced by huge herds of livestock may exceed the ability of the croplands to absorb it, so the rest runs off into the streams that lead eventually to estuaries.

Nutrients also come from the atmosphere—N released from the burning of fossil fuels returns and gets deposited on the land or in the water. The burning of fossil fuels, which emit nitrogen oxides into the atmosphere, initially creates acid rain and air pollution, followed by water pollution once it comes down in precipitation. These nutrients cause algal blooms, followed by hypoxia (low oxygen) in deeper waters, a process called eutrophication (Figure 2.1).

The global rise in eutrophication is due to increases in intensive agriculture, industrial activities, and the human population. There are variations in the importance of each source among regions. For example, in the United States and Europe, agricultural sources (animal manure and fertilizers) are generally the primary contributors, while sewage and industrial discharges (both of which are regulated and usually receive treatment prior to discharge) are a secondary source. Atmospheric sources are also a significant contributor of N in coastal areas. In the Chesapeake Bay, for example, the atmosphere is a major source of all controllable N that enters

The Big Picture

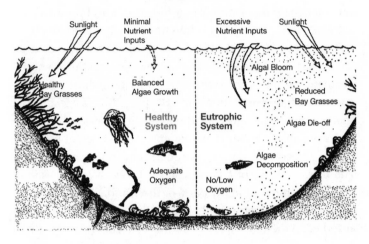

Figure 2.1 Eutrophication (courtesy Chesapeake Bay Program)

the bay. In Latin America, Asia, and Africa, wastewater from sewage and industry are often untreated and may be the primary contributors to eutrophication. In addition to nutrients, wastewater and animal feces also contain harmful microbes that can be removed by sewage treatment plants. Otherwise, these microbes released into waterways may cause disease. This is discussed further in Chapter 10.

How does a sewage treatment plant work?

Sewage treatment removes contaminants from wastewater. The treatment includes physical, chemical, and biological processes to remove physical, chemical, and biological contaminants. The objective is to produce an environmentally safe liquid waste stream (or treated effluent) and a solid waste (or treated sludge) suitable for disposal or reuse (usually as farm fertilizer).

Primary treatment involves the physical separation of solids and liquids. Sewer pipes carry wastewater from homes and industries to the treatment plant. Screens let water pass, but not trash (such as rags, diapers, etc.), which is collected and disposed of. The sewage is then held temporarily in a settling basin where heavy solids settle to the bottom while oil, grease, and lighter solids float to the surface. The settled and floating materials are removed and the remaining liquid may be discharged to the environment (if primary treatment is all there is), or it flows to the next (secondary) stage of treatment.

Secondary treatment involves biological processes by which microorganisms break down organic matter (just like microbes do in the natural environment) in the separated liquid and solid phases. Air is pumped in to the mixture of primary wastewater and microorganisms, whose growth is sped up by the aeration. Final settling basins allow the clumps of microorganisms to settle from the water by gravity. Most of this mixture, called activated sludge, is returned to the aeration basins to maintain the needed amount of microorganisms. Secondary treatment may include a separation process to remove the microorganisms from the treated water prior to discharge or tertiary treatment. The final effluent (liquid portion) may be disinfected (e.g., with chlorine, ozone, UV) before it is discharged, to reduce the microorganisms in the water before it is released into the environment.

Some sewage treatment plants include tertiary processes to remove more pollutants before the effluent is released. In some cases it is pumped to constructed wetlands (with cattails or other plants) for further treatment. Plants in treatment wetlands take up some of the remaining nutrients from the effluent. Other forms of tertiary treatment use biological nutrient removal technology to remove most of the organic N by converting it to N_2 gas, which is harmless and released into the atmosphere (which is mostly N_2 anyway). Effluents must meet EPA's National Pollutant Discharge Elimination System (NPDES) criteria. Improving wastewater treatment systems is a major way to improve water quality.

The sludge from primary treatment is pumped to a separator, where inorganic solids (grit) are separated from the lighter weight organic solids, which are then concentrated and pumped to the anaerobic digesters with bacteria that work in the absence of oxygen. Stabilized sludge has little odor. Methane gas is produced by this anaerobic digestion and can be used as fuel. The final sludge (biosolids) can be used in an environmentally acceptable manner as a fertilizer and soil conditioner except in urban areas, where the sludge may contain high concentrations of metals and other toxic pollutants from industry and must be disposed of. In the past, it was barged out to certain sites in the ocean and dumped, but due to its negative impacts, dumping became illegal. (Virtually all material dumped in the ocean in the United States today is sediments dredged from the bottom of water bodies in order to maintain navigation channels.)

What is Combined Sewer Overflow (CSO)?

Older cities like New York and those in Europe have combined sewer systems built in the nineteenth century that combine sewer pipes from buildings with pipes and sewers for stormwater from streets. This was a major advance from the cesspools and gutters that formerly carried waste to nearby waterways. In the first half of the twentieth century, many sewage treatment plants were designed and built. At that time, combined sewers seemed like a good way to carry away stormwater along with the garbage, animal waste, and other refuse that collected on city streets. Sewage plants were designed to handle twice the average flow of wastewater, but because of population increases, what was considered excess capacity is no longer adequate after a storm. While the volume of water entering sewage treatment plants can be managed during dry weather, during storms the increased volume of water from the storm sewers combined with the wastewater is more than the plants can process. Many sewer systems were designed to accommodate a so-called five-year storm—a rainfall so extreme that it is expected to occur, on average, only twice a decade. But in 2007 alone, New York City had three 25-year storms—storms so strong they would be expected only four times each century. Severe storms are likely to intensify with the forecasted climate change. When treatment plants are swamped, the excess water spills from overflow pipes. To avoid wastewater backing up into homes or streets, outlets allow untreated water, including untreated feces and industrial waste, to be released directly into the waterways (such as the infamous Gowanus Canal in Brooklyn). When a treatment plant releases untreated waste, it is breaking the law. Sewage systems are frequent violators of the Clean Water Act. In the past several years, over 9,000 of the nation's 25,000 sewage systems—including those in major cities—have released untreated or partly treated human waste, chemicals, and other hazardous materials into water bodies. The raw sewage ruins the water quality, including at nearby bathing beaches. Hundreds of older municipalities with combined sewer systems face the same water quality problems during major rainstorms; however, the cost of replacing these systems with systems that separate wastewater from stormwater is very high. Combined sewer overflows have become a major source of pollution, and the resulting bacterial contamination from the sewage can become concentrated in shellfish, rendering them unfit for consumption. Holding tanks and additional treatment plants have been built to cope with the overflow, but permanently correcting the CSO problem will take a sustained policy and many billions of dollars.

What are Concentrated Animal Feeding Operations (CAFOs)?

Concentrated Animal Feeding Operations (CAFOs) with dense concentrations of farm animals generate tons of manure containing nutrients; pathogens, including bacteria and viruses; sediment; antibiotics; and metals, such as copper or arsenic. While farms can apply manure to crops, the amount of manure that CAFOs generate often exceeds local needs. The problem of excess manure—associated with large animal facilities—is found in many areas. Chesapeake Bay is an estuary with major inputs from industrial-scale chicken farms on the Eastern shore. The EPA is charged with protecting our waters from such pollutants, but its regulations have not kept pace with the rapid growth of CAFOs.

What effects do excess nutrients have, or what is eutrophication?

Nutrient enrichment of marine waters promotes excessive growth of algae, both attached multicellular forms such as sea lettuce (*Ulva*) and microscopic phytoplankton blooms. Small increases in algae can increase productivity in food webs and sustain more fish and shellfish. However, overstimulation of algal growth can severely degrade water quality and threaten human health and living resources. When algal blooms eventually die off the dead cells sink to the bottom, where they stimulate bacteria to decompose them. The decomposition process uses up dissolved oxygen from the water. If the aeration of water by mixing is less than the oxygen used up by bacterial metabolism, the bottom waters will become hypoxic (low oxygen) or anoxic (no oxygen), creating stressful or lethal conditions for bottom dwellers. Hypoxia is a major problem in many estuaries, especially in late summer and early fall, and has been increasing globally.

Zones of low oxygen reduce the abundance and diversity of adult fish and reduce the growth rate of newly settled lobsters, crabs, and juvenile flounder. Blue marlin, other billfish, and tunas are rapid swimmers that need high levels of dissolved oxygen, and the expansion of hypoxic areas shrinks the useable habitat for these valuable fishes. Species that cannot move or move slowly may die in low-oxygen zones; disease resistance can be compromised, and reproduction and embryonic development can be impaired. Fish larvae are poor swimmers, and become more vulnerable to predation. In general, animals attempt to cope with low oxygen by reducing their activity in order to consume less oxygen. This often means feeding for shorter periods of time and eating less food. When bottom water is hypoxic, buried clams move up closer to the surface of the sediments and are more easily eaten by blue crabs that are more tolerant of the low oxygen and can make brief trips into the hypoxic zone. Marine worms that have high tolerance for low oxygen show sublethal effects—they have lower respiration and feeding rates, and the fertilization and development of their embryos are abnormal.

Sometimes in warm weather, crabs and other animals swarm into shallow water and may actually crawl out of oxygen-depleted water as they try to breathe. This phenomenon has been referred to as a jubilee. This may result in the crabs surviving hypoxia long enough to be caught for dinner by humans, who gave the name jubilee to this event (although the crabs are clearly not happy).

What effects are seen in seagrasses?

Phytoplankton blooms make the water more turbid, reducing the light available to submerged aquatic vegetation (seagrasses) on the bottom. Seagrasses are an important component of the ecosystem, and they have been in decline. They provide a nursery habitat and play an important role in ecosystem structure and function. They are damaged by both the shading of light and sulfide toxicity from eutrophication. The shading by dense phytoplankton blooms reduces their ability to photosynthesize. Submerged aquatic vegetation (SAV) such as eel grass also declines because of the growth of small algae (called epiphytes) attached on its grass blades. Epiphytes cover the blades with a layer of fuzz and further reduce the amount of sunlight that can reach leaves for photosynthesis.

Epiphytes can be controlled to some degree by small animals like amphipods that graze on them. Grazing by these small animals can be important in keeping seagrass beds healthy, and the more diverse the grazers are the better they clean the epiphytes off the blades. SAV declines usually involve sudden decreases in abundance rather than gradual changes, and high salinity and temperature intensify the effects. Seagrasses also suffer from blooms of sea lettuce, which causes reduction in their shoot density, leaf growth, and carbon content. Seagrass and the detritus it generates provide food and shelter for a variety of animals, and when its growth is reduced the associated animal community declines. Sadly, there are only a few cases of seagrass recovery following the reduction of nutrient inputs.

What effects are seen in coral reefs?

Agricultural runoff with nutrients and sediments is transported to coral reefs by river discharge. Eutrophication is especially harmful to coral reefs, where the nutrients stimulate benthic algae to grow over, cover, and smother the corals, eventually leading to the replacement of the coral reef community with an algal community—especially when grazers (e.g., sea urchins, parrotfish) are not plentiful. Only if the reef has large populations of grazing herbivores to control the algae can the corals survive and prevail in this competition. All too often, these grazing herbivores are reduced due to fishing, and reefs get covered with algae. Reefs off the Florida Keys are degraded because of wastes from too many people, while parts of the Great Barrier Reef in Australia suffer from agricultural runoff. Degradation represents a major loss to tourism, since divers prefer to be in areas with rich coral reef environments.

What is a dead zone?

In many areas hypoxia is so severe that the areas are referred to as "dead zones" because nothing (aside from bacteria) can live there. When the dissolved oxygen (DO) declines below 0.5 mg/l, mass mortality occurs. Areas with oxygen sufficient to sustain some life (below 2 or 3 mg/l) have reduced benthic communities, comprised of very small animals. When the benthic community is stressed by low DO, only short-lived, small surface deposit-feeding worms remain; other animals like crustaceans, bivalves, and gastropods can no longer survive. Increasing numbers of dead zones have been reported globally, a result of runoff and nitrogen deposition from burning fossil fuels. Certain species of phytoplankton including tiny forms (e.g., Cyanobacteria) are favored over diatoms, which are more important in the food web. About 150 dead zones have been identified around the world, including a very large one in the Gulf of Mexico that receives water from the Mississippi River, which drains much of the agricultural center of the United States. This watershed encompasses 41% of the contiguous United States and contains a large portion of the nation's agricultural land. While hypoxia has been noted in the past, it did not become widespread until the 1960s. It tends to be overlooked until fish kills occur and benthic fisheries collapse. Ecosystems that experience long periods of hypoxia have low levels of benthic fauna and productivity of fishes. As documented by Nancy Rabalais and her colleagues from Louisiana Universities Marine Consortium, every summer the oxygen in the Gulf of Mexico dead zone drops to less than 0.5 mg/l (ppm)—conditions under which hardly anything other than microbes can live. The high nutrients from the runoff lead to an overgrowth of phytoplankton that causes the formation of the hypoxic water mass, which lasts from spring through late summer annually. Hypoxic conditions have become more severe since the 1950s as the N input from the Mississippi River into the Gulf has tripled. This hypoxic zone threatens valuable commercial and recreational Gulf fisheries. The size of the 2012 dead zone was the fourth smallest since monitoring began in 1985, at about 2,889 square miles. This was due to the drought that reduced the amount of runoff. In contrast, in 2011, flood conditions, carrying large amounts of nutrients, resulted in a dead zone measuring 6,770 square miles, the size of the state of New Jersey. In 2013 it was up to 5,840 square miles—a bit bigger than Connecticut.

Dead zones now affect more than 400 systems, and cover vast areas of the ocean—more than 475,000 square kilometers (183.4 sq miles). While trends show increases worldwide, some localized areas are improving. Of the 415 areas around the world identified as experiencing some form of eutrophication, 169 are hypoxic and only 13 are classified as being in recovery. In Chesapeake Bay the dead zone affects the distribution and abundance of fishes including croaker, white perch, spot, striped bass, and summer flounder, which are key parts of the ecosystem and support commercial and recreational fisheries. Scientists saw a drastic decline in species richness, species diversity, and catch rate under low dissolved oxygen conditions, suggesting that the fishes begin to avoid an area when levels of DO drop below about 4 milligrams per liter, as they start to suffer physiological stress. The response at this value is interesting because it is greater than the 2 mg/l that is the formal definition of hypoxia. Efforts to reduce inputs of fertilizers, animal waste, and other pollutants into Chesapeake Bay appear to be helping. The size of summer dead zones in deep channels of this bay has been declining.

Can excess nutrients damage salt marshes?

Coastal wetlands support fisheries, protect coasts from storms, and provide habitat for wildlife. They are also able to absorb nutrients from runoff, thereby protecting the nearby estuaries. Salt marshes have been disintegrating and dying over the past two decades along the US Eastern seaboard and other developed coastlines for unknown reasons. Small-scale experiments have shown that nutrient addition results in a decrease in the ratio of plant roots to shoots, an increase in above-ground tissues, a change in plant species composition, and increased vulnerability of plants to herbivores. Linda Deegan and a team of scientists working on entire tidal creeks added nutrients to the water coming into a marsh on rising tides, similar to nutrient-enriched estuaries, and compared the responses with control creeks over nine years. As expected, plants grew smaller roots because nutrients were easier to find, and the decomposition of organic matter in the soil increased because the extra nutrients enabled bacteria to break it down more easily. But other results were unexpected. After a few years, cracks formed in the banks of the high-nutrient creeks, which then collapsed down into the creeks—eventually turning the vegetated marsh into a mudflat, which is less productive and does not provide equivalent habitat for fish and wildlife. The loss of roots and organic matter reduced the stability of the creek-bank soils, leading to the collapse of creek banks and the eventual conversion of salt marsh into mudflat. These results demonstrate the value of long-term field studies and show that salt marshes have a finite ability to absorb nutrients before they themselves are damaged.

How widespread is eutrophication?

Reports of marine and coastal hypoxic areas or dead zones have been increasing in recent years, with increased population growth, urbanization, and expansion of agriculture. The World Resources Institute has compiled maps and identified 415 areas around the world that are experiencing eutrophication, but there are many areas where there is not enough information to determine the extent of eutrophication or identify sources of the nutrients. The increase in reports of hypoxic areas may also be due in part to more scientists looking for them. Many areas in the United States and Europe are improving as a result of environmental legislation in the 1970s. An interactive map is available at http://www.wri.org/project/eutrophication/map.

What are Harmful Algal Blooms (HABs)?

Some phytoplankton species, usually dinoflagellates, produce toxins that can impair respiratory, nervous, and other functions and even cause the death of fish, shellfish, seabirds, and mammals. HABs have been called red tides or brown tides because of water discoloration when they occur, though many kinds of harmful algae do not discolor the water. Their economic impacts can be severe if shellfish harvest and fishing are closed. Reports of HABs have been increasing worldwide, and some correlations have been shown with N inputs. There are examples from around the world where increases in nutrient loading have been linked to the development of large blooms with toxic effects. Not only has the frequency of reports of HABs been increasing, new toxin-producing species have been discovered to cause problems. However, attempts to relate trends in HABs to nutrients are difficult because of variability in geographical regions and over time. They are also complicated by other factors, including increased monitoring and reporting and the influence of climate change. Evidence of a link in one region should not be considered evidence of a general linkage of HABs to nutrients everywhere. Possible causes for the apparent expansion of HABs include natural dispersal of species by currents and storms and dispersal through human activities such as shipping, shellfish translocation, and eutrophication. There are also more scientists out there looking for HABs. Some aspects of the global expansion of HABs could also be due to improved detection of HABs in places where toxic species have always been present and were not stimulated by the human activities.

What are some harmful algal species?

HABs have occurred periodically along the coast of southwest Florida for over a century, causing the death of many species including turtles, manatees, dolphins, and crabs. Decreased abundance of shrimp and several fish species have also been noted, and shellfish farms have been forced to shut down. Mass mortality of birds after eating fish that had consumed toxic algae has been reported. The microscopic dinoflagellate that causes these blooms (*Karenia brevis*) produces a powerful toxin, brevetoxin, which paralyzes animals that ingest it. It not only kills fish, but this lipid-soluble toxin can biomagnify up the food chain to top carnivores like dolphins. It can also cause skin irritation and burning eyes among swimmers; people who are not even in the water may cough and sneeze when winds blow its toxic aerosol onshore. Along the Gulf Coast, *K. brevis* blooms directly affect human health. Eating shellfish with brevetoxins causes neurotoxic shellfish poisoning. In addition, brevetoxin levels in dead fish and fish-eating birds collected from beaches and rehabilitation centers during blooms suggest that brevetoxin can cause bird mortality. These blooms are stimulated when seasonal changes in wind patterns move nutrients east from the Mississippi River. The N-rich river water spurs the growth of the algae, which are pushed by winds toward Florida, concentrating them into larger blooms. In the spring of 2013 a record number of manatee deaths (more than 500) was attributed to this HAB, plus other causes. Scientists thought that toxins in the bloom settled onto the seagrasses that manatees eat, causing them to become paralyzed and eventually to drown. Bottlenose dolphins are also vulnerable to the toxin. In 1999–2000, 152 dolphins in the area died following extensive *K. brevis* blooms; brevetoxin was detected in 52% of the animals. Dolphin stomachs frequently contained brevetoxin-contaminated menhaden fish. In 2005–2006, 90 bottlenose dolphins died when there were high densities of *K. brevis*. Most (93%) of them tested positive for brevetoxin.

In New England, the dinoflagellate *Alexandrium* produces a toxin, saxitoxin, which accumulates in mussels and clams that consume phytoplankton. Humans who eat the shellfish can become seriously ill with paralytic shellfish poisoning (PSP). Toxins from algae can transfer through the marine food web as well, sometimes with a lethal impact on fish or marine mammals.

Diatoms of the genus *Pseudo-nitzschia* produce domoic acid, the cause of amnesic shellfish poisoning (ASP). Blooms of *Pseudo-nitzschia* are common in Monterey Bay, California, causing sea lion mortality. *Pseudo-nitzschia* and domoic acid have been detected in the open ocean in addition to fjords, gulfs, and bays, demonstrating their presence in diverse environments. The toxin has been measured in zooplankton, shellfish, crustaceans, echinoderms, worms, marine mammals, birds, and sediments, which shows how it transfers through the marine food web.

In the 1990s a very bizarre organism, a dinoflagellate called *Pfiesteria* (nicknamed the "cell from hell") showed up in the waters of North Carolina and Chesapeake Bay, producing open sores and killing billions of fish (mostly menhaden) and, frighteningly, causing neurological symptoms in the investigators who were studying it. Researchers found that its toxins cause neurological symptoms including memory loss, disorientation, and speech impediments. One researcher had to be hospitalized before adequate laboratory precautions were worked out. *Pfiesteria* spends much of its life as harmless-looking microscopic cysts in the sediment. But when large numbers of fish are present under the right conditions, it goes through a remarkable transformation in which the cysts turn into toxic flagellated vegetative cells, move toward the fish, release a toxin that kills the fish, and then transform themselves into large amoebae that eat the dead fish. When finished feeding, they revert to their cyst form and return to the sediment. Other stages of this remarkable organism are able to photosynthesize using chloroplasts that they retain from algae that they ate previously.

Pfiesteria was a major problem in water with high nutrient levels from sewage and agricultural runoff. It made headlines and caused considerable concern and controversy in the scientific community for many years. Part of the controversy was due to an inability to isolate a toxin from the cells. The toxin was finally identified chemically, and was found to be a very unstable chemical that disappears from the water quickly, which accounts for the difficulty in finding it. *Pfiesteria* toxicity varies from nontoxic to highly toxic. Toxic strains are capable of killing fish by both toxins and physical attack from feeding upon the skin. Some strains do not produce enough toxin to kill fish, but can kill fish larvae by physical attack. From 1991 to 1998 *Pfiesteria* was linked to major kills of juvenile Atlantic menhaden (*Brevoortia tyrannus*) in the two largest estuaries on the US mainland, but for some reason, it hasn't been heard from lately.

In the 1980s a brown tide of tiny organisms, *Aureococcus anophagefferens*, in eastern Long Island had severe impacts on eel grass populations and the once-thriving Peconic Estuary bay scallop industry, estimated at one time to be worth $2 million. This population has not recovered after many years during which there have been only occasional reoccurrences of moderate brown tides.

How widespread is their occurrence?

HABs have been known throughout history and their incidence appears to be increasing, but there is also greater awareness and research into the problem. The varieties of toxic algae include many species, and HABs have been appearing more frequently around the world. This increase includes more frequent blooms of familiar species as well as blooms of new species not previously known to be harmful or not known at all. Over the past two decades many more toxic species have been identified. The number of reported outbreaks of PSP increased from fewer than 20 in 1970 to more than 100 in 2009. However, some of the apparent increases may be due to increased surveillance.

Efforts are under way to improve detection, so it can be done in real time directly in the ocean. An optical sensor called the BrevBuster can measure *K. brevis* and beam the information back to shore via satellite. A technology that could be used for more algal species is the environmental sample processor (ESP), a robotic lab that collects water samples, extracts and sequences DNA, and transmits the results back to shore. It will provide an accurate, fast, and cheap method to detect HAB cells and their toxins.

What can be done to reduce farm runoff?

Inputs from agriculture can be reduced by using certain types of tillage that reduce runoff and planting buffer strips or trees along stream edges to absorb runoff. There are various farming techniques that reduce runoff, and incentive programs have been set up to encourage farmers to adopt them. Boards placed in farm ditches can block the water in the ditch from entering the drains. This both reduces the amount of water going into streams and slows down the water, giving the N more time to convert to N_2 gas, which dissipates. Reducing fertilizer use and recycling manure will remove sources of nutrients. Unlike point sources of pollution, runoff control and reduction is largely voluntary. There is funding available from the US Department of Agriculture's Conservation Reserve Program, which encourages farmers to reduce erosion and runoff from their farms. Large animal-feeding operations (CAFOs)—the huge sheds containing hogs and chickens—are supposed to be regulated as point sources under the Clean Water Act like factories, but they are not. To control nutrients in runoff, farmers can implement nutrient management plans, which optimize crop yields while protecting the environment. These plans identify the correct timing and amounts of fertilizer that should be applied to fields, thereby reducing the chance for it to be carried away in runoff. Farmers can use manure or litter generated by their cows or poultry as fertilizer for their fields. However, the amount of manure and litter generated on a farm may exceed the amount that is needed for the crops. To help farmers avoid stockpiling extra manure, some states support nutrient transport programs that export it to farms where the nutrients are needed.

Some groups are working to find new uses for manure which might be used in energy production or sold as fertilizer. Some creative approaches include anaerobic digesters that convert the methane from cow manure into electricity that runs the farm and produces extra electricity that can be sold to the power company to provide electricity for nearby homes. This approach not only reduces methane emissions (which contribute to global warming) and turns a waste into a resource, but it also eliminates runoff.

The state of Pennsylvania is trying to get farmers to reduce runoff by letting them apply for pollution credits that can be sold to developers to build sewage treatment plants. Pennsylvania has asked farmers to build barriers to reduce runoff into Chesapeake Bay and, with monetary incentives, to plant crops year-round so that the roots will prevent the soil from washing away in big storms. The state will estimate how much pollution has been eliminated, using an equation that combines the impact of the improvement and the distance of the farm from the bay.

Much effort and a huge amount of money has been expended over the years in attempts to improve water clarity, seagrass populations, and oxygen levels in Chesapeake Bay, but the results have been disappointing. Every year, people wade into the bay to see how deep they can go and still see their feet, a test of water clarity. The deeper they can go, the clearer the water. Unfortunately there has not been much improvement as the number of people in the watershed continues to increase.

In order for significant progress to be made, more cooperation from farmers and substantial financial assistance are
necessary, along with upgrades to sewage treatment plants.
Changes in land-use practices are required in states that are
hundreds of miles away from the affected estuaries. Some
individual farmers are changing their practices to reduce pollution, which may also improve the quality of their topsoil and
sustainability of their farm. There is some good news: since
2006, farmland with cover crops increased from 12% of acres
to 52% in the Chesapeake Bay region. Farmers are using a variety of other conservation practices, such as no-till, that help
keep nutrients and sediment on farm fields and out of nearby
waterways. Some form of erosion control has been adopted on
97% of cropland acres in the watershed.

We do not yet understand how much reduction in nutrient
inputs is needed to produce the needed improvement in water
quality, and what the time lag will be before improvements are
seen. Voluntary efforts to control nonpoint runoff have been
encouraged for two decades, but they don't seem to be able to
deal with the magnitude of the problem.

For areas with impaired water quality, the EPA has outlined
steps needed for pollution reduction called the total maximum
daily load (TMDL), which calls for states to reduce nutrients
flowing into their estuaries, using data and modeling to calculate how much reduction is called for. Much needs to be done
to help state and local authorities address eutrophication, and
the federal government is providing information and technical assistance. A systematic, nationwide plan is necessary to
make real progress in reducing the damage to coastal areas,
and making sure that no other healthy areas become affected.

What can be done to reduce runoff from cities and suburbs?

In urban areas, Clean Water Act regulations require operators of stormwater systems to implement stormwater management programs. Stormwater runoff is a top issue in urban

areas where water bodies are no longer fishable and swimmable (goals of the Clean Water Act). Stormwater not only adds nutrients, but also contributes to flooding in low-lying coastal areas and cities, which have miles of impervious surfaces that cannot absorb the water. Urban inputs can be controlled by reducing the amount of impervious surfaces, sweeping litter off streets before it gets into the water, and improving sewage treatment plants. New York City, for instance, has instituted zoning laws requiring new parking lots to include landscaped areas to absorb rainwater, established a tax credit for green roofs with absorbent vegetation, and begun environmentally friendly infrastructure projects. Philadelphia is building rain gardens and sidewalks of porous pavement and planting thousands of trees. Rain barrels and rain gardens are the subject of educational programs for numerous communities and school groups. These programs provide homeowners with information on installation and maintenance and typically include a hands-on training where homeowners install a rain garden in a community. These are all excellent ways of chipping away at the problem. But unless cities require developers to build in ways to minimize runoff, the volume of rain flowing into sewers is likely to grow.

What can be done about combined sewer overflow?

Some cities have built retention basins—tanks that hold sewage until the water volume following a storm decreases. Other municipalities have reconfigured treatment facilities to expand and maximize flow rate. Still others have adapted green infrastructure—green roofs, porous pavements and bioswales, or planted ditches that filter contaminants—to reduce the amount, speed, and toxicity of water drainage after a storm. A new technology, inflatable dams, has recently been installed in two locations in Brooklyn, NY. These large cylindrical rubber structures are placed within sewer mains and inflate during heavy rain to block the flow of rain water and sewage. They turn the sewer mains into wastewater storage sites; however, if the water level gets too high and threatens to back up into homes or streets, sensors deflate the dam to release some water. Each dam can retain about two million gallons of water until the rain decreases and the dam deflates to allow water to flow to the treatment plant. The dams are expected to save about 100 million gallons of sewage from flowing untreated into the harbor each year.

What techniques in the water can reduce effects of eutrophication?

Marinas can provide pumpout facilities for boats so that they do not discharge their wastes into the water. Techniques to absorb N once it reaches the water are also possible using biology—seaweed farms will absorb nutrients; culturing oysters, clams, or other bivalves will consume large amounts of phytoplankton, reducing eutrophication and at the same time providing food. Oysters are powerful filter feeders that can clear the water as they feed. One adult oyster can filter and remove nutrients from 1.5 gallons of water in an hour. Oyster populations have declined greatly along the East Coast of North America in the past century from overharvesting, pollution, and diseases. It is estimated that one hundred years ago Chesapeake Bay was clear, because the oysters filtered it every three weeks as opposed to every three years today. To help overcome this loss of oysters, planting of oyster reefs has become a very popular restoration procedure. Many pounds of N can be removed by oyster reefs through the process of denitrification by associated bacteria—which returns the N to the air in the form of N_2 gas. In the Chesapeake, scientists found that one acre of oyster reef could remove 543 lbs of N in a year, 25% more than intertidal sediments without oysters. Oyster reefs not only reduce eutrophication but also provide habitat for many other organisms (at a site in Chesapeake, 24,000 organisms were living on one square meter of oyster reef!).

Shrimp, blue crabs, gobies, blennies, and many other animals live on oyster reefs. Oyster reefs also serve to reduce wave and storm surge impacts in coastal areas and support fishing, since some commercially important fish are more abundant in oyster reefs than in nearby mudflats.

Coastal wetlands can also absorb a lot of the nutrients in runoff. There are many projects restoring salt marshes for the numerous services they provide, of which absorbing nutrients is only one. (But as the Deegan study discussed earlier showed, marshes have their limits before they, too, are damaged by excess nutrients.) Marshes also protect the coastline from storm surges, and provide habitat and food for many marine animals and terrestrial ones like shore birds.

What is the prognosis for eutrophication in the future?

While coastal ecosystems may recover when nutrient inputs are reduced, it is a very slow process. Currently, hypoxia is among the most harmful human influences in the marine environment. Although there has been legislation in Europe, levels have not been improving, except in Danish waters. Where eutrophication has been reduced it has generally been from improvements in point sources (sewage treatment) rather than reducing nonpoint runoff or atmospheric deposition, which are much more difficult to control. There has been little progress in reducing nonpoint sources. Global river nutrient export has increased steadily since 1970, with South Asia accounting for at least half of the increase. Under various future scenarios, nutrient exports could change significantly over the next 30 years. Eutrophication is likely to continue to impact freshwater and coastal ecosystems into the foreseeable future.

3

MARINE DEBRIS

Why is marine debris so abundant?

Marine debris is any solid manufactured item that enters the marine environment, including cigarette butts, fishing line, diapers, bottles and cans, syringes, and tires. It is a pervasive pollution problem that has been made worse by the increasing use of plastics, which are the most common constituent of the debris. The UN estimated that 6.5 million tons reaches the ocean yearly—roughly 17,000 tons every day. Over 4.5 trillion cigarette butts are discarded annually and are not only unsightly, their constituents (e.g., nicotine) are toxic to marine life. Plastic debris in the oceans is now so common that even very remote beaches have plastics washed up on them. Depending on the weight and size, marine debris may float, but most of the litter sinks to the seabed.

Where does marine debris come from?

Land-based sources are responsible for 80% of marine debris, after being blown into the water or coming from creeks or rivers or storm drains. Sources include sewer overflows, solid waste (landfills), and litter from streets. Improper control of solid waste in many countries is responsible for much of the debris, which may enter the water directly or indirectly. Other debris comes from ships, recreational boats, offshore drilling rigs, and fishing piers. Materials can be dumped, swept, or blown off vessels and platforms, or can result from littering, dumping in rivers and streams, and spillage of materials during production or transportation. Commercial fishing is a major source of lost nets and ropes. Derelict vessels sit on the bottom of ports and waterways, creating a threat to navigation. Many sink at moorings, or remain partly submerged in the intertidal zone or stranded on the shoreline. One unusual source of a great deal of debris, large and small, was the 2011 earthquake and tsunami in Japan, which sent houses, docks, cars, and everything they contained adrift in the Pacific Ocean.

What are the major constituents of debris?

Plastics, as mentioned before, comprise a large proportion of the debris, and the variety and quantity of plastic items has increased dramatically, including domestic material (shopping bags, cups, bottles, bottle caps, food wrappers, balloons) (Figure 3.1), industrial products, and lost or discarded fishing gear. As these materials are commonly used, they are common in marine debris. Derelict fishing gear includes nets, lines, crab and shrimp pots, and other recreational or commercial fishing equipment that has been lost, abandoned, or discarded in the water. Modern gear is generally made of synthetic materials and metal, so lost gear can persist for a very long time. Monofilament fishing line can persist for hundreds of years. Glass, metal, and rubber are used for a wide range of products. While they can be worn away—broken down into smaller and smaller fragments—they generally do not biodegrade entirely. Today, most of what we use comes packaged in plastic, which can last for centuries. It is this stability and resistance to degradation that causes it to be so problematic. A generation ago, products were packaged in reusable or recyclable materials like glass and paper. Today, we use products that we dispose of at the end of their short life, and which end up in landfills, on our beaches, and in the ocean as marine litter.

In addition to the visible litter on beaches, microscopic plastic debris from many sources including the breakdown of

Figure 3.1 Marine debris on a beach (photo from NOAA)

larger pieces as well as residue from washing synthetic fabrics is accumulating in the marine environment and could be entering the food chain. Researchers traced some microplastics back to synthetic clothes, which release thousands of tiny fibers when they are washed. Microplastic is abundant on shorelines, especially near urban areas, and consists of polyester, acrylic, and polyamides (nylon). Litter also accumulates in the deep sea, not an eyesore to us, but a problem for deep sea animals.

What happens to the plastic? Does it break down?

Plastic is extremely slow to degrade and tends to be buoyant, which allows it to travel in ocean currents for thousands of miles. Most plastics become brittle when exposed to ultraviolet (UV) light and break down into smaller and smaller pieces, forming microplastic. These pieces, as well as plastic pellets, are already found in most beaches around the world. No one knows just how small these pieces become—they are very difficult to measure once they are small enough to pass through the nets typically used to collect them. Their impacts on the marine environment and food webs are still poorly understood. These tiny particles are known to be eaten by various animals and to get into the food chain. Due to its low density, plastic waste is readily transported long distances and concentrates in gyres, which are systems of rotating ocean currents. We don't know how long plastic remains in the ocean. Current research suggests that most commonly used plastics will never fully degrade. Because most of the plastic in the ocean is in very small fragments, there is no practical way to clean it up. One would have to filter enormous amounts of water to collect a relatively small amount of plastic.

How is debris in the ocean measured?

The most common way to measure floating plastic in the ocean is to collect it using very fine-meshed nets towed at the ocean surface from a ship. These nets collect planktonic organisms, as well as plastic and any other floating debris, which is sorted to pick, count, and preserve all plastic samples collected during the tow. However, a lot of the debris floats below the surface and is not collected by towing nets at the surface.

How much is there?

Annual cleanups pick up millions of pounds, mostly plastic, from beaches, although most beaches around the world are not cleaned up. We don't know how much trash is out there because no one monitors it carefully. There are also large quantities of small debris mixed in the sand or within the water column. The Ocean Conservancy, a Washington, DC-based environmental organization that organizes cleanups, released its 2012 list of trash collected during its International Coastal Cleanup. More than 10 million pounds of debris was collected by volunteers globally, with over 769,000 pounds from California alone.

The top ten items found during the cleanup were (1) cigarettes/filters, 2,117,931; (2) food wrappers/containers, 1,140,222; (3) plastic beverage bottles, 1,065,171; (4) plastic bags, 1,019,902; (5) caps/lids, 958,893; (6) cups, plates, forks, knives, spoons, 692,767; (7) straws/stirrers, 611,048; (8) glass beverage bottles, 521,730; (9) beverage cans, 339,875; and (10) paper bags, 298,332. Of course, cigarettes would comprise a much smaller fraction if the amounts were calculated by weight or volume.

The Cayman Islands' main tourist attraction is its marine life and beaches. From far away these beaches look beautiful, but during an international coastal clean-up, in one and a half hours volunteers filled 98 garbage bags with about 1,500 pounds of trash from just six miles of Caymans' beaches. Most of the waste was, not surprisingly, plastic, with bottles and containers accounting for over 50% of the waste. Of the 12 seas surveyed by the United Nations Environment Programme (UNEP) between 2005 and 2007, the Southeast Pacific, North Pacific, East Asian Sea, and Caribbean coasts contained the most litter, and the Caspian, Mediterranean, and Red Seas had the least. Studies of the Baltic Sea, Northeast Atlantic, the United States coastline, and the North Atlantic Subtropical Gyre indicated no major changes in the amount of litter between 1986 and 2008. Within the United States, however, litter increased from 1997–2007.

The actual amounts are far greater than estimated from surveys of floating litter, since much of the heavier material sinks, and much of the lighter material is pushed downward in the water by currents. Winds blow light pieces of plastic down below the surface, causing researchers to greatly underestimate the amount of plastic.

Why does debris accumulate in large patches in the middle of the ocean?

Marine debris that does not accumulate along shorelines can be blown by the wind or follow the flow of ocean currents, often ending up in the middle of oceanic gyres (circular current patterns) where currents are weakest. The ocean water is constantly moving, carrying water, organisms, and debris around the globe. As material is captured in the currents, wind-driven surface currents gradually move floating debris toward the center, trapping it in the region. Flotsam from San Francisco can reach the North Pacific Gyre in as little as six months. Crab trap tags and floats lost from the state of Oregon during 2006–2007 were recovered four years later in the Northwestern Hawaiian Islands. The Great Pacific Garbage Patch is a vast region of the North Pacific Ocean. Estimated to be double the size of Texas, the area contains over 3 million tons of plastic, mostly in small pieces. In this area waste material from across the North Pacific, including coastal North America and Japan, are drawn together. Contrary to what its name implies, the area is not a concentration of trash visible in satellite or aerial photographs. There is not a giant island of solid garbage floating in the Pacific. Rather, there are millions of small and microscopic pieces of plastic floating over a roughly 5,000 square km area of the Pacific. The amount has increased significantly over the past 40 years, and plastic debris there apparently already outweighs zooplankton by a factor of 36 to one. Islands within the gyre frequently have their coastlines covered by litter that washes ashore—prime examples being Midway and Hawaii, where plankton tows sometimes come up with many more plastic pieces than plankton. The next largest known marine garbage patch is the North Atlantic Garbage Patch, estimated to be some hundreds of kilometers across. There are other smaller patches in the Southern Hemisphere.

Where else does debris accumulate?

Litter can end up anywhere. Not all of it floats; some of it is heavy and sinks out of sight. Scientists at the Monterey Bay Aquarium Research Institute (MBARI) analyzed 18,000 hours of underwater video collected by remotely operated vehicles

(ROVs) on the bottom of the deep Monterey Canyon. In this region, researchers noted over 1,150 pieces of debris on the seafloor. About one-third was plastic objects, and of these more than half were plastic bags. Metal was the second most common type of debris; about two-thirds were cans of aluminum or steel. Other common debris included rope, fishing equipment, glass bottles, paper, and cloth. The trash was concentrated on steep rocky slopes. Surprisingly, it was common in the deeper parts of the canyon, below 2,000 meters (6,500 feet). In the same areas where trash accumulated, there was also wood and other natural debris that originated on land, leading researchers to conclude that much of the trash came from land-based sources rather than ships. Previous studies underestimated the extent of marine debris in the deep due to lack of technology for observing deep bottoms. In another study, microplastics were found in remote deep-sea sediments collected at locations ranging in depth from 1,100 to 5,000 meters.

Researchers collecting samples in the Southern Ocean that encircles Antarctica have detected high levels of plastic pollution in an area that was considered unspoiled. Similarly, the sea bed in the Arctic deep sea is becoming covered with litter. Photographs taken to investigate biodiversity of sea life provided evidence of increasing debris. Waste, primarily plastic, was seen in 1% of the images from 2002, but in the images from 2011 it had doubled. While 2% does not seem like much, the deep-sea Arctic Ocean has been considered to be one of the most remote and pristine parts of the oceans.

Granted it is ugly, but can the litter harm marine life?

Effects of marine litter are primarily physical rather than chemical. Debris that washes in and covers salt marshes or mangroves injures the plants, the base of the food web in these tidal wetlands. Injuries and subsequent recovery depend on the extent and type of debris. Marine debris affects animals through ingesting it or getting entangled in it; it is estimated that up to 100,000 marine mammals, including endangered species, are killed each year by marine debris. Very serious effects happen when marine animals become entangled in debris such as fishing line and six-pack rings. Birds get fishing line entangled around their legs, which get injured and may be lost. Large amounts of plastic debris have been found in the habitat of endangered Hawaiian monk seals, including in areas that serve as nurseries. Entanglement in plastic debris has led to injury and deaths in endangered Steller sea lions, with packing bands the most common entangling material. Hatchling sea turtles run down the beach to the ocean, a critical phase in their life cycle. Debris can be a major impediment if they get entangled in fishing nets or trapped in containers such as plastic cups and open canisters. Marine debris is an aspect of habitat quality for nesting sites and may help explain declines in turtle nest numbers on certain beaches. Many marine birds such as Northern Gannets use plastics as nesting material. Gannet nests studied contained an average of 470 grams of plastic, which translates to an estimated colony total of 18.46 tons. Most of the plastic used was synthetic rope. About 63 birds were entangled each year at one study site, totaling 525 individuals over eight years, the majority of which were nestlings.

Many marine animals consume flotsam by mistake, as it often looks similar to their natural prey. Sea turtles, for example, may mistake plastic bags or balloons for jellyfish, a favorite food. At study of stranded sea turtles in Australia found that larger individuals had a strong preference for soft, clear plastic, lending support to the idea that they ingest debris that resembles jellyfish. Smaller turtles were less selective in their feeding, though they tended to prefer rubber items such as balloons. Young sea turtles in the western Atlantic have a stage described as their "lost year," when they are thought to live among the floating seaweed *Sargassum*. Juveniles collected from *Sargassum* have been found to ingest plastic debris that floats along with the seaweed. Plastics in diet samples averaged 13%, suggesting that the *Sargassum* habitat comes with a risk of ingesting much debris. A study of fishes from the general region of the North Atlantic gyre found extensive marine debris ingestion in seven species with 58% in one species (*Lampris* sp., small-eye). Of all sampled individuals, 19% contained some debris, mostly plastic or fishing line. Surprisingly, species that ingested the most debris are ones considered to live in intermediate depths rather than near the surface and therefore unlikely to come into contact with surface debris, suggesting that there is more debris below the surface than we thought.

It is difficult to prove that a dead animal died from ingesting debris, however. In 2008 two sperm whales were stranded along the California coast with large amounts of fishing net scraps, rope, and other plastic debris in their stomachs. Plastic debris that becomes lodged in digestive tracts, blocking the passage of food, can cause death through starvation. A sperm whale that beached itself in 2012 in Spain had a large amount of garbage blocking its stomach, including some 36 square yards of plastic canvas, a dozen meters of plastic rope, plastic sheeting used on the outside of greenhouses, plastic sheeting used inside, and even two flower pots. The whale was emaciated because its intestines were totally blocked by the plastic debris. A Risso's dolphin in the Hudson River in May 2013 likely starved to death because of four plastic bags lodged in its stomach. The dolphin, which was 10 feet long and weighed 600 pounds, had four intact plastic bags in its stomach. The largest bag was 4 feet by 2 feet and was rolled into a sphere, about 8 inches in diameter, blocking its stomach.

Debris is also ingested by marine birds, which may starve or become strangled if an object becomes lodged in their throats or digestive tracts. Ingested marine debris is commonly found in dead birds, turtles, and other animals, although one cannot assume the debris caused the death (Figure 3.2). Scientists quantified the stomach contents of 67 Northern Fulmars from beaches in the eastern North Pacific in 2009–2010 and found

Figure 3.2 Bird carcass with ingested plastic (photo from NOAA)

that 92.5% of the birds had ingested an average of 36.8 pieces, or 0.385 g of plastic. Compared to earlier studies, this shows an increase in plastic ingestion over the past 40 years. New approaches allow the study of stomach contents in living birds by giving them ipecac, so that they vomit. Almost half the storm petrels sampled in Newfoundland had ingested plastic. Many adult seabirds feed ingested plastic to their offspring, so chicks likely have a higher plastic burden than their parents.

Tiny floating microplastic particles also resemble zooplankton, so they can be eaten by filter feeders and enter the food chain. Approximately 35% of filter-feeding fish studied near the North Pacific Gyre had ingested plastic in their stomachs, averaging 2.1 pieces per fish. Catfish studied in Brazilian estuaries had plastic in their stomachs; 18% of the individuals of one catfish species and 33% of the individuals of the other species. All developmental stages (juveniles, subadults, and adults) were contaminated. Nylon fragments from fishery activities were the major constituent. Plastic contamination was high in Norway lobsters; 83% of the animals sampled contained plastics (mostly filaments) in their stomachs. Tightly tangled balls of plastic strands were found in 62% of the animals. Some of the microfilaments in the gut contents could be traced back to fishing waste.

To add insult to injury, chemical pollutants like DDT and PCBs (described in Chapter 6) collect on the surface of plastic debris, thus making the plastic a source of toxicity, transferring chemicals into the food web where they can then accumulate in birds and other marine animals that eat the plastic. Plastic debris was collected in the North Pacific Gyre and analyzed for polychlorinated biphenyls (PCBs), organochlorine pesticides like DDT, and polycyclic aromatic hydrocarbons (PAHs) (see Chapter 6). Over 50% of the plastic contained PCBs, 40% contained pesticides, and nearly 80% contained PAHs. The concentrations of pollutants found ranged from a few parts per billion (ppb) to thousands of ppb. The types of PCBs and PAHs found were similar to those found in marine sediments. In addition to these chemicals on their surface, marine plastics contain additives such as plasticizers, antioxidants, antistatic agents, and flame retardants. Some additives (e.g., nonylphenol, bisphenol A) cause endocrine disruption—they interfere with body processes mediated by hormones. This can result in impaired nervous system development, abnormalities in behavior, malformations, and disruption of normal sexual development and reproduction.

Consuming plastic is an entry point for contaminants that were either initially a constituent of the plastic, or gathered from the water, into the marine food web. Plastic debris can become more toxic as bigger pieces break up into smaller pieces, increasing the surface area available for gathering pollutants. The smaller the debris, the greater the likelihood it can be ingested and introduce contaminants into the small organisms low on the food web. Mark Browne and colleagues found that toxic concentrations of pollutants and additives enter the tissues of animals after eating microplastic. They exposed lugworms (*Arenicola marina*) to sand with 5% microplastic that also contained common chemical pollutants (nonylphenol, phenanthrene) and additives (triclosan, PBDE-47) and showed that the pollutants and additives from ingested microplastic were present in the worms' tissues at concentrations that can cause harmful effects. A companion paper by Stephanie Wright and colleagues found that when there is a lot of microplastic in the sand, lugworms eat less and have reduced energy levels. Lugworms are common benthic burrowers that can comprised up to one-third of the mass of benthic organisms on some shores; they churn up the sediments like earthworms on land and are consumed by birds and fish and used as bait by fishermen.

Worldwide, over 250 species are known to become entangled or to ingest marine debris and an estimated 100,000 marine mammals and turtles are killed annually by litter. Debris can damage habitats if it covers coral reefs, marshes, or seagrass beds. When plastic film settles to the bottom, it can suffocate immobile plants and animals or wrap around corals.

One piece of good news for one species is that the floating litter provides a habitat for marine insects called sea skaters. These relatives of pond water skaters live at the surface of the water and lay their eggs on floating objects. They are able to use the plastic garbage as a new site for depositing their eggs, leading to an increase in egg densities in the North Pacific Subtropical Gyre. Small pieces of plastic also provide habitat for multitudes of microbes colonizing and thriving on flecks of plastic—a new human-made environment for microbial communities that has been termed the "plastisphere."

What problems are caused by derelict fishing gear?

Lost fishing nets (made of nondegradable plastic) and traps may settle onto the bottom and continue to trap fish for years (this is known as ghost fishing). Lost fishing gear catches ecologically and economically important animals, including protected and endangered species. It is estimated that about 5,000 crab pots are lost annually from the Alaskan crab fishery alone. Abandoned or lost fishing gear can be a navigation hazard and have significant economic impacts. In a study of lost gillnets in Puget Sound, Washington, scientists estimated the daily catch rate of a single lost gillnet, which still catches crabs, and developed a model to predict overall mortality. They calculated that over 4,000 Dungeness crabs would be entangled during the lifetime of a single derelict net, which is a loss of over $19,000 to the fishery, compared to a cost of only $1,358 to remove the net. Scientists recovered almost 32,000 derelict blue crab pots from Chesapeake Bay that had trapped 40 different species and over 31,000 organisms. Blue crabs themselves were the most common species in lost pots with an estimated 900,000 killed each year, a potential annual loss to the fishery of $300,000.

Entanglement of the monk seal in Hawaii, an endangered species, is the major impediment to the species' recovery. Monofilament line is single-strand, high-density nylon line on fishing reels. Used line discarded into aquatic environments can damage boat motors and wildlife. Marine animals cannot see discarded monofilament line, so it is easy for them to become entangled and starve, drown, or lose a limb. Illegal driftnets in the Mediterranean Sea are a major hazard to marine mammals, reptiles, and fish. The use of driftnets has been banned by the UN for 20 years and the EU more recently, but an estimated 500 vessels from Morocco, France, Italy, Turkey, Algeria, and Tunisia continue to use them, inadvertently killing whales, dolphins, sharks, and sea turtles.

Australia is home to six of the world's seven threatened species of marine turtles. During a recent cleanup of ghost nets on beaches, 80% of the animals found in trapped in the nets were marine turtles, including Olive Ridley, Hawksbill, Green, and Flatback turtles. Getting tangled in ghost nets is one of the most common causes of death for marine turtles in Australia. Scientists used data on the number of ghost nets found during beach cleanups in the Gulf of Carpentaria and combined that with a model of ocean currents, to simulate the likely paths that nets took to get to their landing spots on beaches. They combined that model with data about where turtles exist in the Gulf, and combined the predictions about where turtles would wash ashore in ghost nets with data on turtle distribution to see where the hotspots are, making it possible to intercept nets before they reach the high-density turtle areas.

What are the biggest pieces of marine litter?

Remnants of buildings and other land-based structures as well as docks were sent adrift by the Japanese earthquake and tsunami of March 2011, along with whole ships that were sent afloat. The Japanese government estimates that approximately 5 million tons of debris washed out to sea. Of that mass, about 1.5 million tons probably floated away and could be transported to the beaches of the northeast Pacific Ocean, including the West Coast of the United States and Canada. Some of these large floating materials are accumulating on the coasts of Hawaii, Oregon, Washington, and Alaska and in the Pacific gyre, a potential human and environmental hazard. Oceanographic studies and models suggest that about 75% of the floating debris will not come ashore. The rest is likely to land along the coast of Alaska over several years. The estimate is that no more than 138,000 tons is likely to land along the coasts of Oregon, Washington, and British Columbia in any single year, with 11,000 tons as a more probable amount. If distributed evenly along the shoreline, this mass of debris would be between 0.5 and 6.7 tons/mile of beach. However, individual beaches may receive significantly more or less accumulation due to local oceanographic effects. The government of Japan has given the United States $5 million in funds to help with clean up of marine debris from the tsunami.

Since hard surfaces are home to many species of attached organisms, this debris can transport many organisms to new places like the Northwest Pacific Islands, where there are some of the best coral reefs in the world, as well as the Pacific Coast of the United States. This issue will be discussed more in Chapter 10, which covers invasive species.

Can marine debris harm people?

Debris can affect human health when broken glass, cans, and medical wastes such as syringes wash up on recreational beaches. Swimmers and divers can become entangled in abandoned netting and fishing lines. Debris that enters the water with sewage (e.g., tampon applicators) may indicate a more serious sewage pollution problem with pathogenic bacteria. Since microplastics pick up toxic pollutants and get into food webs, it is likely that they are getting into our seafood, which is another source of those toxic chemicals for us. Aesthetic problems of ugly litter on the shoreline can easily result in economic effects on coastal communities when tourists stop coming.

What can be done about it? Can cleanups be effective?

Worldwide efforts are underway to monitor and remove marine debris, and to prevent further pollution by controlling litter at its source. Marine debris is a solvable problem if people can identify sources and control them. During the annual International Coastal Cleanup sponsored by the Ocean Conservancy, millions of volunteers in 127 countries around the world pick up debris from beaches and record data. Participants fill out a data sheet to record the specific items collected, indicating the types of activities that produced the litter. This produces an extensive database of information (as well as cleaner beaches) around the world. Every September, volunteers have removed debris from nearly 288,000 kilometers of coast, 60% of which is comprised of fishing lines and nets, beach toys, and food wrappers. Another 29% is cigarette butts and filters. Marine debris items range from 4 to over 48,000 items per kilometer of shoreline. The author participated in a cleanup in January 2009 on a no-smoking beach in Southern California, and plastic, especially Styrofoam, was far more abundant than anything else collected; cigarette butts were extremely rare.

A nonprofit organization called the Rozalia Project has developed an underwater trash robot that uses sonar to detect objects and picks up litter under the water before it can foul the beaches. In 2011 in Boston Harbor, the sonar revealed tires, large pieces of metal, and piles of beer cans and plastic cups (as well as crabs walking along the bottom). It found and picked up over 880 pieces of marine debris from a single pier. This trash robot is expensive, but supported by various corporations.

Fishing for Energy is a partnership designed to provide commercial fishermen with a free disposal solution for unusable fishing gear. The program gives them a convenient place to dispose of gear—bins at busy fishing ports—to eliminate the expense and hassle of disposal of unusable fishing gear in landfills. The program increases the likelihood that derelict gear does not become marine debris. Another example is a multiregional effort to remove abandoned crab traps, which are boating hazards and needlessly trap and kill fish. Monofilament recovery programs have been started in some states in which the fishing public deposits used line in containers. The line is collected and sent to the manufacturer for recycling. The company melts down the line and uses it to manufacture new plastic fishing products such as tackle boxes and spools for lines. New crab pots are being made with a biodegradable panel that will provide an escape for animals trapped in lost pots.

However, the most viable option to reduce litter is to reduce its production in the first place, improve reuse and recycling, and enhance environmental awareness. As a result of public pressure, some plastics manufacturers are responding. When evidence was found that one of the sources is microbeads used as exfoliants in facial scrubs and personal care products, efforts were made by environmental groups to get the manufacturers of these products to stop using the beads and use biodegradable alternatives. Johnson & Johnson agreed to phase out the use of polyethylene microbeads in personal care products such as Neutrogena, and Clean and Clear and has stopped developing new products containing plastic microbeads. Unilever and The Body Shop have also committed to phasing out microbeads by 2015.

California is spending nearly half a billion dollars annually to prevent trash from polluting its beaches, rivers, and ocean. The money is being used by municipalities for river and beach cleanups, street sweeping, the installation of devices to capture stormwater, cleaning and maintaining stormwater drains, cleanup of litter, and public education.

What about public education?

This issue is also amenable to public education and monitoring. Major educational programs and outreach to community groups and schools have been developed, including a YouTube video from cartoonist Jim Toomey (www.youtube.com/watch?v=DtfAhy2lgAA&feature=youtu.be).

An environmental documentary called "Trashed" with Jeremy Irons was released in 2012. A trailer for the movie is on YouTube (www.youtube.com/watch?v=7UM73CEvwMY&feature=youtu.be).

People who fish can be educated to hold on to their waste line and put it in bins instead of tossing it into the water. However, no matter how careful individuals are, tackle can get snagged underwater and retrieving it may be practically impossible.

Are there laws to reduce marine litter?

Rules and regulations can be more effective than voluntary cleanups. Some countries have laws and policies for debris control. There are also international agreements such as the London Dumping Convention and the International Convention for the Prevention of Pollution from Ships (MARPOL). In the United States, the Beaches Environmental Assessment and Coastal Health Act (BEACH) of 2000 was designed to reduce the risk of disease to beach-goers and includes a provision for monitoring and assessment of floatable materials. The federal government is involved in programs to reduce marine litter. NOAA's Marine Debris Program participates in many beach and river cleanups, removals of derelict fishing gear and abandoned boats, but focused most of its energies in 2012 on the Japanese tsunami debris monitoring and cleanup. They also developed a research strategy, standardized methods for monitoring and assessment of marine debris on shorelines and surface waters, and participated in over 100 outreach events, educating nearly 20,000 people about marine litter issues.

Appropriate management of wastes can prevent items such as disposable plastic bags from becoming marine debris. Plastic bags are a major component of litter and are being banned in some areas. The California Coastal Commission found that plastic bags comprise 13.5% of shoreline litter; the City of Los Angeles found that plastic bags make up 25% of the litter in storm drains. Programs are being developed to recycle plastic bags. Recycling of plastic film climbed 4% to reach one billion pounds annually in 2011 for the first time. The category of plastic film includes plastic bags, product wraps, and commercial shrink film. A report developed by Moore Recycling Associates noted that the recycling of plastic film has grown 55% since 2005.

San Francisco and some other municipalities passed ordinances that would ban most retail locations from distributing plastic bags and begin charging customers a dime for each paper bag (or compostable plastic bag) they use. In Toronto, as of January 1, 2013, retailers are prohibited from giving customers single-use plastic bags, including those advertised as compostable, biodegradable, or photodegradable. As of October 1, 2012, the importation, manufacture, or sale of plastic bags and disposable foam products was banned in Haiti. Most such products are currently imported from the Dominican Republic. It is unknown how well this law will be enforced. Unless there are readily available alternatives and consistent enforcement of the ban, it will be ineffective and may well end up hurting some of the people most directly affected by the litter problem when sewer systems back up.

Market-based solutions, such as asking people to pay for plastic bags at checkout, have been effective in Washington, DC and in Ireland. Shoppers in China, Mexico, India, and countries throughout Africa and Europe shop without single-use bags. They bring their own bags to the market. This idea is catching on in the United States, and there are many efforts encouraging shoppers to bring their own bags. Eighth-grade student Emily Miner in Pacific Palisades, California created and sold reusable shopping bags with an historical Pacific Palisades photo. Dan Jacobson, legislative director of Environment California, has included her bag in his collection of creative alternatives to plastic bags. He travels throughout California showing this kind of grassroots effort to reduce single-use bags.

In Louisiana, where marshes are being lost at an alarming rate, recycled plastic will be used to protect restored marshes in Lake Pontchartrain. Seventeen floating islands five feet wide by twenty feet long will be built and placed in front of a man-made marsh. The floating islands, which will be about 18 inches thick, are made from layers of what looks like Brillo pads but are actually recycled plastic bottles. The floating islands are stocked with native plants and microbes and then anchored on the bottom.

An island on the Great Barrier Reef has stopped selling water in plastic bottles to reduce litter. Several other market-based approaches have been explored, such as deposit schemes to encourage the return and multiuse of plastic bottles and taxation on single-use plastics that do not fit into deposit return systems. An organization called the Plastic Bank is setting up plastic repurposing centers around the world in areas with an abundance of both plastic waste and poverty. Their goal is to remove plastic from the environment while helping people rise from poverty. They will provide education and the opportunity to trade reuseable plastics for credits that can be used for microfinance loans or other projects. However, there has been little widespread application of these approaches. Despite the abundance of information, projects, and regulations, marine debris remains a major problem because people still generate the debris and laws are not well enforced. The amount of marine debris is highly variable, but amounts have been increasing by about 5% per year.

Can new technologies reduce the problems of marine debris?

Recycling of plastic is highly effective in some countries (Switzerland, 98%; Germany, 95%), but open landfills and no recycling are still the norm in many places. Ecover, a European cleaning brand, announced that it will use plastic waste from the sea to create a new type of sustainable and recyclable plastic bottle. The company is working to combine plastic waste with a plastic made from sugar cane and recycled plastic for packaging. Boats with trawls will collect plastic waste for cleaning and recycling, while other fishermen will collect plastic debris mixed with bycatch and deposit it at special collection points. The sorted waste will be sent to a recycling plant, where it will be turned into the plastic for the new bottles. A carpet tile company and the Zoological Society of London are cooperating in a program called Net-Works that pays people in the Philippines for used nets, which are recycled into carpet tiles.

The Sea Shepherd Conservation Society is partnering with the Bionic Yarn Co, and Parley for the Oceans in The Vortex Project, which removes plastic from the ocean and transforms it into fashion. The Vortex Project takes waste from the oceans and shorelines, and recycles, enhances, and reuses it for yarn and fabric, for consumer products such as denim. They will also seek to close the loop by again recycling these products at the end of their product life and manufacturing new products in such a way as to not further pollute.

Biodegradable fishing gear can greatly reduce problems caused by ghost fishing. Biodegradable fishing line, based on dissolvable surgery suture technology, has been developed. Modifications are needed to disarm gear once it is lost. Researchers developed an oval panel for fish traps that would degrade, leaving an opening for animals. Once the panel dissolves, any animal that can enter the trap can also escape. Crab pots with a biodegradable panel were tested in the blue crab fishery in Chesapeake Bay. Commercial crabbers deployed pots with the panels as well as standard pots and found no reduction in the catch, showing that biodegradable panels can be a viable solution to derelict pots.

4

OIL AND RELATED CHEMICALS

What are the components of oil?

Oil is a complex combination of various hydrocarbons (carbon-based compounds with hydrogen atoms attached). Petroleum hydrocarbons are the primary constituents in oil, gasoline, diesel, and a variety of solvents. Oil and related substances don't generally mix with water (as you may have heard) but float on the surface, although some lightweight components (the water-soluble fraction) do dissolve, and some evaporate. Hydrocarbons include straight-chain compounds of various lengths, such as octane with eight carbons (Figure 4.1), branched compounds, and compounds in which the carbons form a ring. Aromatic hydrocarbons have alternating double and single bonds between carbon atoms forming a six-part ring. The configuration of six carbon atoms in aromatic compounds is known as a benzene ring, after the simplest possible such hydrocarbon, benzene. Aromatic hydrocarbons can have more than one ring, and are termed polycyclic aromatic hydrocarbons (PAH). The smallest one is naphthalene, with two rings (Figure 4.2).

What are polycyclic aromatic hydrocarbons (PAHs)?

Polycyclic aromatic hydrocarbons (PAHs) are major toxic components of oil. There are thousands of different compounds in this group; the larger molecules (with more rings) are less soluble in water, more soluble in fats, and tend to be more

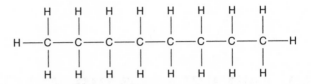

Figure 4.1 Octane molecule, a straight chain hydrocarbon

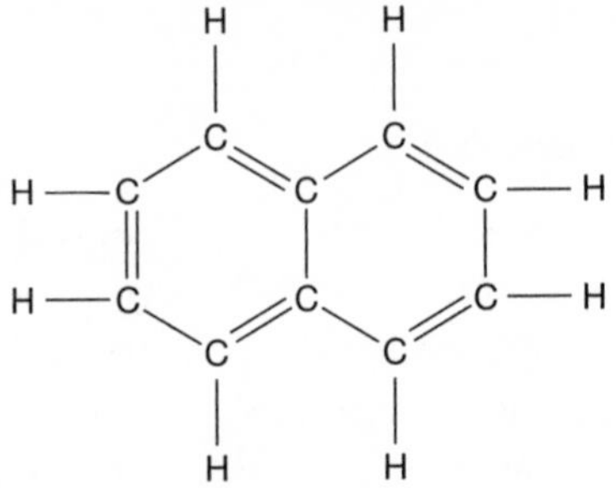

Figure 4.2 Naphthalene molecule, a polycyclic aromatic hydrocarbon

carcinogenic (cancer-causing), mutagenic (causing genetic mutations), or teratogenic (causing embryonic malformations). PAHs are broken down in the liver into compounds that can be excreted, and generally do not increase in concentration as they make their way through the food chain. However, in some cases the intermediate breakdown products can be more toxic than the original PAHs. In aquatic environments PAHs may evaporate or be degraded by bacteria, but they often become more toxic after exposure to light (phototoxicity), and they may become incorporated into sediments where they degrade very slowly. High concentrations in sediments are associated with liver and skin cancers in bottom-dwelling (benthic) fishes. Because mollusks cannot metabolize PAHs, they continue to accumulate them and are thus useful for biomonitoring programs, in which their bodies are analyzed to evaluate the levels of PAH pollution of a particular environment. In addition to oil spills, PAHs enter the marine environment from other sources—they can be released into the atmosphere through burning, and reach aquatic systems in rainfall. Large quantities of PAHs are also released directly into the water by chemicals leaching from creosote-treated wood that is used for bulkheads and docks in shallow water. There was an interesting case in which an unfortunate female herring laid her eggs on a creosote bulkhead in California, and all the eggs were highly abnormal or dead. This particular bulkhead was over forty years old and releasing far less creosote than it did when it was new, but it still was lethal to all the embryos, as Carol Vines and coinvestigators at University of California, Davis found. PAHs also are found in runoff from asphalt-paved surfaces. Parking lots and roads may be sealed with PAH-containing coal-tar sealants, and, especially after the paving has recently been treated, large amounts of PAHs leach out into runoff whenever it rains.

PAHs are emitted into the air when fossil fuels are burned. They come down into the ocean in precipitation, and can affect marine life. Corals are particularly sensitive to these aerosol particles, which contain soot (carbon with attached PAHs) and sulfates from fossil fuels.

What are the major sources of oil in the ocean?

Petroleum hydrocarbons (including both linear compounds and PAHs) have been a long-standing problem in the marine environment. There is great public concern about oil spills and the resultant shoreline fouling and deaths of large numbers of marine birds and mammals. Oil spills can be spectacular and do a considerable amount of damage; the most obvious effects being seen in marine birds and mammals that live at the surface or must come up to the surface to breathe, while other marine organisms lower in the water may suffer chronic effects. Since oil spreads as a thin layer on the surface, spills can pollute large areas and represent serious environmental hazards. The major source of accidental oil input into seas is from tankers and pipelines (about 70%), while the contribution of offshore drilling is lower. Most spills are relatively small and result from routine operations such as loading and discharging in ports or oil terminals. The largest spills arise from accidents (groundings and collisions) involving tankers carrying large amounts of oil. Catastrophic spills that release more than 30,000 tons of oil are relatively rare but can cause the most serious ecological damage (primarily for sea birds and mammals) and produce long-term environmental damage as well as major economic impacts. The highly publicized 1989 spill of the *Exxon Valdez* in Prince William Sound, Alaska caused unprecedented damage to the fragile Arctic ecosystem. The other major oil disaster in recent years is the blowout of the mile-deep BP well Deepwater Horizon in the Gulf of Mexico, which gushed oil continually for four months.

What happens to the oil after it is spilled?

Floating oil has its greatest effects on animals that live at the sea surface such as birds and sea otters. Over time the lighter components evaporate, leaving behind the heavier components. When the oil comes into shallow water and coastal marshes it can coat and smother the resident communities. If it arrives on rocky shores and coats rocks, the lighter components of the oil evaporate, leaving behind the heaviest components and turning the oil into tar that covers the rocks but eventually erodes away due to wave action. Under those circumstances, biological communities will return rather rapidly since the oil is removed relatively quickly. However, when the oil reaches soft substrates like salt marshes or beaches, it can sink below the surface and persist for many decades.

What happened with the Exxon Valdez?

On March 24, 1989, the *Exxon Valdez* oil tanker struck Bligh Reef in Prince William Sound, Alaska and spilled 11 million gallons (equivalent to 17 olympic-sized swimming pools) of crude oil. The ship had encountered icebergs in the shipping lanes, and Captain Hazelwood ordered the helmsman to go out of the shipping lanes to go around the ice. He then handed over control to the third mate with instructions to turn back into the shipping lanes when the tanker reached a certain point. They failed to make the turn back into the shipping lanes soon enough, and the ship ran aground. For the first few days after the spill, most of the oil was concentrated in a large patch near Bligh Island, but no effective cleanup had started. On March 26, a storm with winds over 70 mph weathered much of the oil, thickening it into mousse and tar balls and distributing it over a large area. By March 30, the oil extended 90 miles from the spill site. The spill occurred at a time of year when the tidal fluctuations were nearly 18 feet, causing the oil to spread onto shorelines way above the normal zone of wave action. The oil eventually covered 1,300 miles of coastline, and 11,000 square miles of ocean. Of the 1,300 miles of coastline, 200 miles were heavily or moderately oiled (the impact was obvious) and 1,100 miles were lightly or very lightly oiled (light sheen or occasional tar balls). This spill was the largest ever in US waters at that time in terms of volume released. In addition, Prince William Sound's remote location, accessible only by helicopter, plane, or boat, made government and industry response efforts difficult and slow.

What were the causes of the accident?

The National Transportation Safety Board investigated the accident and found that the probable causes of the grounding were:

1. The failure of the third mate to properly maneuver the vessel, possibly due to fatigue and excessive workload;
2. The failure of the master to provide a proper navigation watch, possibly due to impairment from alcohol;

3. The failure of Exxon Shipping Company to supervise the master and provide a rested and sufficient crew for the ship;

4. The failure of the US Coast Guard to provide an effective vessel traffic system;

5. The lack of effective pilot and escort services.

What actions were taken after the spill to protect shorelines?

Five trials of dispersants (which are like detergents, making the oil soluble in water) took place between March 25 and March 28, but by March 29 the regional response team decided that dispersants were no longer feasible. Federal, state, and local agencies decided that fish hatcheries and salmon streams had the highest priority. Containment booms were deployed as physical barriers to protect these areas; about 100 miles of boom was deployed. Almost all the types of boom were used and tested. Due to the extent of the spill it was necessary to employ inexperienced workers to deploy and tend booms, and this led to some booms being incorrectly used or handled, and sometimes damaged. The primary means of open water oil recovery was with skimmers. In general, most skimmers became less effective once the oil had spread, emulsified, and mixed with debris. Sorbents were used to recover oil in cases where mechanical means were not feasible. The use of sorbents was labor intensive and generated large quantities of solid waste. Sorbent booms were used to collect sheen between layers of offshore boom, and from the beach during tidal flooding.

How was the spill cleaned up?

It took more than four summers of cleanup efforts before efforts were ended. At the height of the response, more than 11,000 personnel, 1,400 vessels, and 85 aircraft were involved in the cleanup. Not all beaches were cleaned, and some beaches remain oiled today. It is believed that wave action from winter storms did more to clean the beaches (moving the oil elsewhere) than all the human effort involved. Exxon says it spent about $2.1 billion on the cleanup effort. A number of techniques were attempted. Beach applications of dispersants were tried in several locations. Corexit® 7664 was applied on Ingot Island, followed by a warm water wash. No significant change in oil cover or the physical state of the oil was observed as a result of this treatment, however. High-pressure cold water treatment and hot water treatment involved dozens of people spraying the beaches with fire hoses. The water, with floating oil, would trickle down the shore and be trapped within several layers of boom and either be scooped up, sucked up, or absorbed using special oil-absorbent materials. Hot water treatment was stopped after it was found that it seemed to be causing more damage than the oil by effectively cooking small organisms in the sand. Mechanical cleanup was attempted on some beaches using backhoes and other heavy equipment to till the beaches, exposing the oil underneath which could be washed out. Many beaches were fertilized to promote growth of bacteria that metabolize or break down the hydrocarbons. This type of bioremediation was extensive in 1990, with 378 shoreline segments fertilized to promote bioremediation. Monitoring of the effectiveness of bioremediation on over 20 beaches determined that oil degradation had been enhanced, but some disagreement existed over whether bioremediation was solely, or even largely, responsible. This method of bioremediation was successful on beaches where the oil was not too thick. A few solvents and chemical agents were used, although none extensively. An important observation was that natural cleaning processes were often very effective at degrading the oil. It took longer for some sections of shoreline to recover from invasive cleaning methods (hot water flushing in particular) than from the oiling itself.

There is still a considerable amount of oil remaining in sediments; fifteen years later some fish and wildlife species injured by the spill had not fully recovered. It is less clear, however, what role oil played in their inability to bounce back. An ecosystem is dynamic and continues its natural cycles and fluctuations while responding to oil. As time passes, it becomes more difficult to separate natural changes from oil-spill impacts.

Have there been some resulting policy changes to prevent future spills?

The US Coast Guard now monitors tankers via satellite as they pass through Valdez Narrows, cruise by Bligh Island, and exit Prince William Sound. In 1989, the Coast Guard watched the tankers only through Valdez Narrows and Valdez Arm. In 1990, the US Congress enacted legislation requiring that all tankers in Prince William Sound be double-hulled by the year 2015. It is estimated that if the *Exxon Valdez* had had a double-hull, the amount of oil spilled would have been reduced by more than half.

What happened with the well blowout in the Gulf of Mexico?

On April 20, 2010, the Macondo well blowout occurred approximately five thousand feet below the surface of the Gulf of Mexico, causing the BP-Transocean drilling platform Deepwater Horizon to explode, killing eleven workers and injuring others. About five million barrels of crude oil were released into the sea—on average, sixty thousand barrels a day (about 11,350 tons of gas and oil per day) before the gusher was finally capped on July 15. Over 630 miles of Gulf Coast shoreline were oiled, mostly in Louisiana.

What responses were taken?

There were over 400 controlled burns, which killed hundreds of sea turtles and unknown numbers of dolphins and other animals. To protect marshes from incoming oil, booms were set around islands and shorelines, and two million gallons of the dispersant Corexit® were applied on and beneath the surface of the sea to break up the oil. Dispersants are complex mixtures of chemicals that have surfactant (wetting) properties, which allows them to act as emulsifiers, essentially letting the oil and the water mix. After extensive use, oil was no longer visible on the surface of the water, and some claimed that it was gone and degraded by microbes.

Why was the use of dispersants so controversial?

By enhancing the amount of oil that physically mixes into the water column, dispersants reduce the amount of oil that reaches the shoreline. Dispersants also stimulate the natural process of aerobic biodegradation by breaking oil up into tiny droplets that are so diluted that the natural levels of available nitrogen, phosphorus, and oxygen are sufficient for microbial growth to degrade the oil. On the other hand, once the oil is dispersed in deep water, it cannot be recovered. When combined with dispersants in the water, oil may be more toxic than either the oil or the dispersant alone. Most studies found that the combination of oil and dispersant increased toxicity. Two dispersants, Corexit® 9500 and 9527A, were used. Although they are EPA-approved, they are ranked by EPA as more toxic and less effective than other dispersants.

What happened to the oil and dispersants?

The well blowout occurred in deep water, where a turbulent discharge of hot pressurized oil and gas mixed with seawater and dispersed by itself into droplets, emulsions and gas hydrates without the use of chemicals. This naturally dispersed mixture did not rise to float on the surface as oil typically does but stayed in a subsurface plume. Amid reports of the oil being nearly "all gone," a plume of hydrocarbons about 22 miles long in deep water over 3,000 feet below the surface was discovered, residue from the well blowout. The continuous plume of oil persisted for months without substantial biodegradation. Dissolved oxygen concentrations suggest that microbial respiration in the plume was low. The dispersants created dispersed oil plumes in deep water because the high pressures and low temperatures made the mixture of oil, dispersants, seawater, and methane neutrally buoyant. Subsequently, it was found that much of the dispersant itself was contained in the plume in the deep ocean and had still not degraded three months after it was applied; it seemed to have become trapped in deepwater plumes. The toxicity of this mixture on deep sea communities is unknown, as are the impacts on planktonic filter feeders and fish eggs and larvae in the water column. Eventually microbial activity degraded the oil. If dispersants had not been used, the surface oil might have been weathered into tar balls by the time it reached the coast. This would have created a public relations nightmare on beaches and affected the socioeconomic activities of residents and tourists. The dispersed oil below the ocean surface appears to have killed benthic animals in intertidal and shallow subtidal regions on and near sandy beaches. In the wetlands only the fringe-edge marsh plants were damaged by the toxic oil and dispersants, since these plants appear to have absorbed the chemicals that caused the death of shoots.

The Deepwater Horizon blowout was unprecedented because of the use of dispersants at the wellhead, resulting in subsurface retention of oil as finely dispersed droplets and emulsions and deepwater retention of plumes of natural gas that underwent rapid microbial degradation. Eventually natural oil-degrading bacteria worked on the plumes and rapid degradation took place, despite the low 5°C temperature. This took place in the deep water, as a result of the geography of the Gulf of Mexico, which is fairly enclosed. When the hydrocarbons were released from the well, bacteria bloomed and then swirled around in the currents and often came back repeatedly over the leaking well. Thus, water that already had a bacterial community got a second input of hydrocarbons, and the microorganisms that had already bloomed and degraded hydrocarbons immediately attacked and degraded the new oil.

In addition to the one-fourth of the oil that was degraded, the Unified Command, led by the US Coast Guard, physically removed about a third of it, and burning at the surface removed another 5%. However, the oil budget they published was criticized as incomplete. Samantha Joye of the University of Georgia said that data she collected showed that oil at depth, as well as gas, lingered much longer than the oil budget suggested. There was also the residual oil unaccounted for, some of which is still out there—on or under beaches, in marshes, down on the bottom, or floating as tar balls.

What were the overall impacts to the ecology of the Gulf?

At the time this book is being written, mostly in 2012 and 2013 there have been relatively few published reports on effects, as many scientists are not permitted to publish their findings and it is too early to say anything about long-term effects on the Gulf ecosystems. Three years after the accident, fish in the Gulf were found with high levels of petroleum hydrocarbons in their tissues, presumably because dispersants made them more available. The floating brown alga *Sargassum* is an oasis of biodiversity and productivity, functioning as habitat for a diverse collection of attached and mobile animals. The Deepwater Horizon oil contacted much of the Gulf of Mexico's *Sargassum*. Aerial surveys during and after the spill showed loss and subsequent recovery of the seaweed. Dispersant and dispersed oil caused it to sink and reduced the local dissolved oxygen. *Sargassum* accumulated oil, exposing animals to toxicants; application of dispersant sank the *Sargassum*, thus removing the habitat and transporting oil and dispersant deeper; and low oxygen levels around the algae stressed the resident animals.

Coral larvae were damaged by the oil, and even more so by the dispersants. Scientists also found that dispersants enabled the oil to penetrate more deeply into sand on the seabed, where low oxygen levels would slow down its degradation. Severe injury was seen in some deep sea corals. The presence of recently damaged and dead corals beneath the known path of a plume from the well is strong evidence that the oil damaged deepwater ecosystems. It is too early to assess the overall impacts of this disaster, but a committee of the National Academy of Sciences (NAS) recommended an ecosystem services-based evaluation. The people who live and work in the Gulf region depend on ecosystems for services such as food and fuel, flood and storm protection, and tourism and recreation. Damage to natural resources from the oil spill could impair these services, leading to social and economic impacts that may not be apparent from an assessment of environmental damage alone. The NAS committee introduced an ecosystem services approach that requires understanding environmental impacts from a disruption, the resulting decrease in goods and services, and the cost of those losses to individual communities and society at large. They illustrated how this approach might be applied to coastal wetlands, fisheries, marine mammals, and the deep sea—each of which provide key ecosystem services in the Gulf.

What happens when oil reaches shorelines?

When spilled oil reaches a rocky shoreline, lighter components of the oil evaporate, leaving behind the heavier components and turning the oil into tar, which will erode away due to wave action, and biological communities will return rather quickly. In marshes, however, oil can sink below the surface and remain for many years. Oil accumulated in marsh sediments undergoes some microbial breakdown, but the process is very slow, and marshes have the slowest rates of recovery from oil spills. Marshes and sediments in Prince William

Sound in Alaska retained oil for many years after the massive oil spill of the *Exxon Valdez* in 1989. What remains controversial is how long the effects persisted. In a cold environment like Alaska oil degrades much more slowly than in warmer regions, and salmon embryos in the sediments a decade later did not develop properly. After over a decade pockets of oil remained in these marshes, and mussels, clams, harlequin ducks, and other birds that feed on sediment-dwelling invertebrates showed evidence of harm in some localized areas. Fish embryos continued to be affected by oil trapped in gravel and sediments for many years after the spill, according to Ronald Heintz and colleagues from NOAA.

Knowledge about long-term consequences of spilled oil should be included when assessing oil-impacted areas. It will take many years to understand the overall impacts of the enormous oil pollution that gushed into the Gulf of Mexico from the Deepwater Horizon in the spring and summer of 2010.

How does oil harm marine birds and mammals?

There are three primary ways oil injures wildlife: (1) it coats the fur and feathers and destroys the insulation causing the animals to die of hypothermia (they get too cold); (2) animals eat the oil, either while trying to clean it off their fur and feathers or while scavenging, and the oil is toxic; (3) oil impairs them in long-term chronic ways, such as damaging the liver or impairing reproduction. An impaired animal cannot compete for food and avoid predators.

What kinds of toxic effects does oil produce in other marine animals?

Corals

An oil spill in Panama initially caused coral bleaching (symbiotic algae are expelled from coral tissue), tissue swelling, mucus production, smaller gonads, and dead areas on corals, even in reefs that had not been in direct contact with the oil. Hydrocarbons in reef sediments were correlated with the degree of injury and with reduced growth. The probable cause of high injury was chronic exposure to sediments mixed with partially degraded oil that had moved from mangroves onto reefs. There was no evidence of recovery five years after the spill. Years later, reduced colony size and decreased size of gonads were seen, which can reduce the number of reproductive colonies. Oil damages the coral reproductive system resulting in fewer breeding colonies, fewer ovaries per animal, fewer larvae, premature release of larvae, abnormal behavior of larvae, and lower growth rates. Dispersants appear to increase the damage done by the oil.

Fish and Crabs

Oil has major impacts on fish embryos that may produce delayed effects when they become adults. Pink salmon that had been exposed as embryos to *Exxon Valdez* oil and survived to migrate to the ocean, returned from the sea at only half the normal rate. These adults showed reproductive impairment and their embryos had reduced survival. Thus, the second generation was affected by the exposures their parents had had as embryos. Oil spill effects on fish eggs have been intensively studied, with studies initially focusing on the water-soluble fraction (WSF) containing mostly one- and two-ringed aromatic hydrocarbons. After the *Exxon Valdez* spill, which occurred during the breeding season of many fish, fish embryos were exposed to partially weathered oil including the larger three-, four-, and five-ringed hydrocarbons, which had been thought to be less toxic than the WSF. These PAHs affected pink salmon and herring eggs at concentrations far lower than had been previously known to be toxic. Ronald Heintz and colleagues found more deformities and chromosomal abnormalities in embryos from oiled than from unoiled locations, even years after the accident. Furthermore, the toxicity of some oils to fish embryos is greatly increased by light (phototoxicity). John Incardona and colleagues studied Pacific herring embryos following the *Cosco Busan* spill in San Francisco Bay and found that components of the oil accumulated in embryos, then interacted with sunlight at low tide to kill them. Three months after the spill, embryos caged at deeper sites in oiled areas had sublethal heart toxicity (as expected from exposure to PAHs), but intertidal embryos that were exposed to oil in the light had very high rates of mortality. The toxicity of dispersed oil tends to be higher than that of crude oil.

After the Deepwater Horizon blowout, Gulf killifish embryos exposed to sediments from areas that had been oiled in 2010 and 2011 showed developmental abnormalities including heart defects, delayed hatching, and reduced hatching success. Killifish are abundant in the coastal marsh habitats and are important members of the ecological community. Because they are nonmigratory, measurements of their health reflect their local environment.

Atlantic bluefin tuna and other large fishes spawn in the vicinity of the Deepwater Horizon, raising the possibility that their eggs and larvae, which float near the surface, were damaged by the oil. The developing hearts of tuna were found by John Incardona and colleagues to be damaged by oil, which interrupts the ability of the heart cells to beat effectively. Oil interferes with heart cell contraction and relaxation—essential for a normal heart beat. Authors concluded that deaths of early life stages of Gulf populations of tunas, amberjack, swordfish, billfish, and others that spawned in oiled surface habitats were likely.

PAH-exposed fish may also develop tumors. Liver tumors in English sole (bottom dwellers, exposed to sediments) in Puget Sound were associated with PAHs in the sediments, as seen by Myers and colleagues from the NOAA Lab in Seattle. The fish with the highest frequencies of liver tumors were from the urban Duwamish Waterway (16%) and Everett Harbor (12%), while sole from other areas had only 0 to 5.5% incidence of tumors. Sediment PAH concentration was correlated with tumors.

Oil is also associated with changes in behavior. Juvenile coho salmon exposed to the water soluble fraction of crude oil showed reduced swimming activity, which was affected by the concentration and time of exposure. When fish were put into clean water, normal activity was restored within eight hours. Awantha Dissanayake and colleagues compared cellular (immune function), physiological (heart activity), and behavioral (feeding) responses in shore crabs collected from a PAH-contaminated site and two cleaner sites and also looked at responses of crabs exposed in the laboratory to pyrene, a four-ring PAH. While no significant cellular or physiological impacts were seen in contaminated crabs, feeding was impaired: when given a cockle (bivalve), the field and laboratory-exposed crabs took longer to handle and break the shells. Therefore, behavior was more sensitive than the cellular and physiological responses.

How long do effects of oil spills last?

Many of the major spills had long-term consequences because the oil came into estuaries and marshes. The persistence of oil is influenced by several factors, such as water solubility, weathering rate, and sediment grain size. The residues may last for decades and continue to affect biological functions including behavior, development, genetics, growth, feeding, and reproduction. Long-term effects have been studied after spills, and they vary depending on the nature of the oil, the temperature, and the nature of the area of the spill. After a spill, most of the oil undergoes a weathering process. However, in marshes or sandy beaches oil can sink down to depths where it persists for decades in the absence of oxygen. The major effects of a rather small oil spill (190,000 gallons of number two fuel oil) in Falmouth, Massachusetts in the late 1960s lasted for over a decade, according to a team of scientists from the nearby Woods Hole Oceanographic Institute led by Howard Sanders, a distinguished benthic ecologist who had been studying the area. It is rare that a spill occurs right in an area that has been intensively studied and was well understood prior to the spill, so their information was particularly useful, though hotly contested by the oil companies. Fiddler crabs were particularly sensitive and were still affected seven years after the spill. The oil affected their burrow construction—the burrows did not go straight down, but leveled off to a horizontal plane, perhaps avoiding the oil below. While this was not a problem during the summer, when winter came the crabs were not deep enough to be below the freezing zone, so they froze to death. Benthic communities took about a decade to return to normal. Over thirty years later, the site of the spill was studied by another generation of Woods Hole scientists, led by Jennifer Culbertson, who found that there was still substantial undegraded oil residue several inches below the marsh surface, and fiddler crab burrows in oiled areas were shorter in length often turned horizontally below 10 cm, sometimes even turning upward. They found that crabs exposed to the oil avoided burrowing into oiled layers, had slower escape responses, reduced feeding rates, and lower population density. Marsh grass in areas that had some oil remaining grew less densely than in clean areas, and the loss of marsh grass (especially roots) made the sediments more likely to erode away. Ribbed mussels were still experiencing effects of the remaining oil. In an experiment, mussels were transplanted from a control site into the oiled site for short-term exposure, and others that had been exposed to the oil were transplanted from the oiled site to the control site. Both the short- and long-term exposure transplants had slower growth, shorter shell lengths, and decreased filtration rates compared to control mussels.

After the *Exxon Valdez* spill the subsurface oil persisted, and chronic exposures continued to affect organisms for over a decade. Three years after the spill most of the remaining oil was in places where it could not degrade, such as below the surface sediments or under mussel beds. Heavily oiled coarse sediments protected oil reservoirs below the surface, preventing oil from weathering in intertidal sites. These sites often contained fish eggs and other vulnerable organisms. In a cold environment like Alaska oil degrades much more slowly than in warmer regions, and salmon embryos developing in the sediments a decade later still did not develop properly. After more than a decade, pockets of oil remained in these marshes where many species continued to show evidence of harm. Fish embryos continued to be affected by oil trapped in gravel and sediments many years after the spill.

Can oiled birds and sea otters be rehabilitated?

Marine birds and mammals are the most obvious victims after spills, since they are large and at the surface of the water. Birds try to clean up oil on their feathers by preening and end up swallowing oil, which is toxic to them, affecting their immune system and making them more vulnerable to disease. Nine years after the *Exxon Valdez* spill, most injured populations had not recovered. Many people have spent a great deal of time and effort to clean oiled birds and marine mammals after oil spills. A study of oiled and rehabilitated brown pelicans found that long-term injury had taken place, and the birds did not breed or show normal behavior or survivability. A study of released oiled, oiled and rehabilitated, and unoiled surf scoters (*Melanitta perspicillata*) after a spill found that scoters tolerated the rehabilitation process itself well, but they subsequently had markedly lower survival than unoiled birds.

How can oil spills be cleaned up?

Considerable expense and effort is associated with attempts to clean up oil spills, which may take months or even years to clean up. Methods for cleaning up include skimming, which requires calm waters and removes the oil quickly without damaging the environment. Bioremediation is the use of microorganisms or biological agents to break down or remove oil. It can be effective, but takes a long time. Controlled burning can effectively reduce the amount of oil in the water if done properly, but it can only be done in low wind and can cause air pollution and kill surface dwelling animals. Dispersants act as detergents, clustering around oil globules and allowing them to dissolve in the water. This improves the surface aesthetically and mobilizes the oil. Smaller oil droplets, scattered by currents, may cause less harm and may degrade more easily. However, the dispersed oil droplets penetrate into deeper water and can affect marine organisms, since they are toxic. Another approach is to just wait; in some cases, natural attenuation of oil may be most appropriate, because of the potential harm associated with some of the methods of remediation, particularly in ecologically sensitive areas. Remedial actions after oil spills are controversial, and some of them (e.g., aggressive cleaning with large heavy equipment) may be worse than the original problem, as was seen with some of the attempted clean up techniques after the *Exxon Valdez* oil spill. However, an oil company advocating for natural attenuation instead of a cleanup would have a major public relations problem.

What are the trends in oil spills over the decades?

The number of spills from tanker ships has decreased greatly over the past three decades. There were about three times as many spills in the 1970s as in the 1990s. However, the number of spills does not consider the volume of oil; the frequency of large spills has decreased as well as the frequency of small ones. Although more oil is being transported in larger supertankers, technical, political, and legal experience in managing the problem has been gained in many countries and internationally through conventions initiated by the International

Maritime Organization (IMO). As a result of the *Exxon Valdez* spill, the United States passed the Oil Spill Act of 1990 requiring all newly built tankers to have a double hull. Any time a tanker is carrying oil it runs the risk of a spill, and now they need to have insurance for unlimited damages—which is good motivation for being very careful. Overall, US oil spillage has decreased over 200% since the 1970s and 150% since the 1980s, despite the *Exxon Valdez* and Deepwater Horizon. International trends are similar. This is encouraging, because there has been an increase in oil transport worldwide in the past two decades. The reduction in oil spills may be due to improved safety standards, and the realization that spills in the United States could result in enormous costs for which the spiller would have unlimited liability according to the Oil Spill Act of 1990. One would predict that oil spills will continue to diminish as there will be less reliance on oil and increased use of alternative energy sources in the future. Furthermore, oil that is transported by sea will be more likely to be contained in double-hulled tankers.

5

METALS

What are the major sources of metal pollutants?

Metals are naturally occurring elements in the earth's crust that can become contaminants when industrial activity concentrates them at higher than normal levels. Since they are elements, they cannot break down into anything else. Metals released from mining and industrial processes are among the major contaminants of concern in coastal environments, where they accumulate in sediments and coastal organisms. Mercury, cadmium, copper, zinc, chromium, and silver are major contaminants from industrial processes, including power plants. Since coal contains mercury, when it is burned the mercury enters the atmosphere where it can be transported long distances before being deposited far from its source. Mercury (Hg) deposited from the atmosphere is a significant fraction of the mercury entering coastal waters and approximately 90% of the Hg in the open ocean. It contaminates seafood commonly eaten by people in the United States and globally. Over the past century, Hg in the surface ocean has more than doubled as a result of human activities. While some metals such as copper (Cu) and zinc (Zn) are essential for life at low concentrations, other metals (Hg for example) play no normal biological role. While most metal contamination originates from land-based industrial sources, metals also are used in antifouling paints for ships. Since fouling (attached) organisms such as barnacles and algae can accumulate on ship bottoms (increasing drag, thus increasing fuel consumption), antifoulant coatings have been developed. For thousands of years ship hulls have been treated with various substances to reduce fouling. Paints containing copper have been used for many years. Beginning in the 1940s organotin compounds (organic chemicals including tin) were developed, and one of the most effective and long-lasting is tributyltin (TBT), which is also one of the most toxic to other organisms.

When marshes were being filled in for development, household and industrial wastes such as metal cans and paint cans with pigments that contain metals were a common component of the fill material. Mercury, cadmium (Cd), lead (Pb), and copper (Cu) from pipes, antifouling paints, and CCA (chromated copper arsenate)-treated wood bulkheads and pilings were common contaminants. While copper is essential for some biological processes and is not generally a concern for human health, it is extremely toxic to algae and invertebrates, and is even used as an algicide and molluscicide.

Lead comes in runoff from road surfaces during rain, from its previous use as a gasoline additive, even though leaded gas is not used any more. Pb remains in the environment and does not break down, so some otherwise fairly pristine marshes have elevated amounts in the sediments as a result of decades of hunting ducks and other waterfowl with lead shot. Lead shot contaminates marsh soils, and birds that normally pick up small pebbles for grit in their digestive system to grind up seeds can consume the spent shot, sometimes resulting in fatal lead poisoning.

Other metals that can be environmental problems include cadmium, chromium, zinc, and copper. Selenium (Se) can be found in different chemical forms and can bioaccumulate in animals and cause deformities under some circumstances, although in other instances it can counteract the toxic effects of mercury. It affects the immune system, alters genes and damages the nervous system, and is particularly toxic to developing embryos. Inorganic mercury and methylmercury (a more toxic form) tend to be more concentrated in sediments with marsh plants than in bare sediments, perhaps due to higher microbial activity in sediments around roots. Common reeds can release a volatile form of inorganic mercury into the air, thereby removing some from contaminated sites but sending it elsewhere.

What are some highly mercury-contaminated sites?

In the late 1950s and 1960s, thousands of people in Minamata, Japan suffered from mercury poisoning. This community had a factory that discharged mercury into Minamata Bay, from which the people ate fish that had accumulated the poison in their tissues. Local residents developed severe neurological and developmental defects, a condition now called Minamata disease, sending a wake-up call to the rest of the world that exposure to mercury can have life-long neurological effects. Thousands suffered from poisoning, which in extreme cases led to insanity, deformation, and death. Many children whose mothers had eaten contaminated fish were born with severe disabilities, even when their mothers had no overt symptoms.

Berry's Creek Marsh, a contaminated Superfund site in the Hackensack Meadowlands of New Jersey, also has extremely high concentrations of mercury in its sediments as a result of industrial pollution. The Hackensack and Passaic Rivers and Newark Bay formed a major center of the Industrial Revolution. Paper, paint, chemical factories, and plants that manufactured gas were some of the early manufacturing facilities in the area, and the factories used the rivers and estuary for wastewater disposal, which at the time was quite legal. As a result, the entire system—not only Berry's Creek—is highly contaminated with PCBs, dioxins, PAHs, and mercury. Fortunately, the mercury has largely not become methylated and has not accumulated to Minamata-type levels in local fish, because of other factors. In this area, there is so much sulfide in the sediments as a result of years of accumulating wastes from sewage treatment plants that the mercury is chemically bound to the sulfide and is not available for bacteria to methylate. Nevertheless, mercury levels in the fish do exceed levels that are considered safe, so fish advisories are posted throughout the system warning people not to consume fish or crabs. While there are other estuaries in the country that are highly contaminated with toxic substances, this one was designated by EPA at the time as the most Hg-contaminated one in the nation.

How does the chemical form of the metal affect what it does?

Knowing the chemical form (species) of a metal is necessary in order to understand its toxicity and the risk it poses. In general for many metals, the free ion—for example, copper with two positive charges (Cu^{2+})—is the most available and toxic form of the metal found in the water. In aquatic environments copper exists in particulate, colloidal, and soluble states, predominantly as metallic (Cu^0) and cupric copper (Cu^{2+}). It forms complexes with both inorganic and organic molecules. The toxicity of Cu is directly related to the free ion, as is the toxicity of Cd, so measurements of total Cu or total Cd in the water overestimate the amount that is bioavailable and potentially toxic. Chromium^{6+} is much more toxic than Cr^{3+}. Organic forms of mercury (e.g., methylmercury) and tin (e.g., tributyltin) are far more toxic than the ionic forms.

Once metals are taken up into an organism, they may be stored in granules within their cells or attached to metal-binding proteins that keep the metals unavailable to the animal (and out of trouble), but metals attached to these proteins can get transferred to a predator that eats the organism. Feeding, or trophic transfer, is the most important way that metals move up the food web into large fish and birds. Plants generally pick up metals from the soil in which they grow, and different species store different proportions in their roots, stems, or leaves, and can pass metals along to animals that consume them.

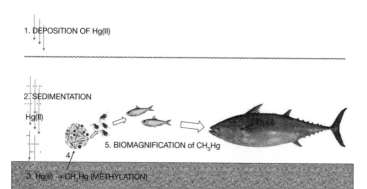

1. DEPOSITION OF Hg(II)

2. SEDIMENTATION

Hg(II)

5. BIOMAGNIFICATION of CH₃Hg

4

3. Hg(II) → CH₃Hg (METHYLATION)

Figure 5.1 Mercury methylation and biomagnification (courtesy Dr. P. Weis)

Although Hg toxicity in highly contaminated areas such as Minamata Bay is well documented, it can also be a threat to the health of people and wildlife in environments that are not so obviously polluted. The risk is determined by the form of mercury and the chemical and biological factors that influence how it moves and changes form in the environment. Inorganic mercury can get transformed into organic mercury compounds. Methylmercury (meHg) is a more toxic form, which is produced from inorganic Hg by the action of bacteria (Figure 5.1). Bacteria capable of methylating Hg^{2+} have been isolated from sediment, water, soil, and fish tissue. MeHg is the form of Hg that is most likely to bioaccumulate in fish and other organisms. MeHg, in addition to being far more toxic than inorganic forms of the metal, also is biomagnified up the food chain, so tissue concentrations increase as it moves up the food chain. Mercury contamination and its health implications are important internationally. Throughout the world, elevated meHg has been found in some fish species that are of economic importance, including shark, swordfish, tilefish, king mackerel, tuna, and Spanish mackerel, as well as freshwater species such as bowfin, largemouth bass, and chain pickerel.

The other organometal of concern is tributyltin, but unlike Hg, tributyltin (TBT) breaks down in the environment, gradually losing its butyl groups over time, reducing its toxicity as it eventually becomes dibutyltin, then monobutyltin and then inorganic tin, which is not toxic at all. Organotins are very toxic, but inorganic tin is not toxic at all.

Where do metals concentrate in the environment?

Metals do not generally reach high concentrations in water but bind to sediment particles, from which they are available to varying degrees to marine organisms, particularly benthic (bottom-dwelling) ones, from which the metals move up the food chain. Bioavailability of sediment-bound metals is a critical issue for their toxicity. Since smaller sediment particles have more surface area for binding metals, the fine particles of silt and clay in estuaries and marshes bind more metal than sand, resulting in high contaminant levels in the fine sediments that are consumed by some benthic animals. These bottom-dwellers can also absorb the metal from the water surrounding the sediment particles, known as pore water, while others directly eat the sediments. Acid volatile sulfide (AVS) in sediments binds metals and has been used to predict the toxicity in sediments of some metals, including copper (Cu), cadmium (Cd), nickel (Ni), lead (Pb), mercury (Hg), and zinc (Zn). The rationale is that the AVS in sediment reacts with the metal to form an insoluble metal sulfide that is relatively unavailable for uptake. Estuarine sediments tend to have high levels of sulfide (as in Berry's Creek), and thus relatively low bioavailability of sediment-bound metals. Ironically, elevating the oxygen in overlying water decreases AVS, thereby increasing metal availability. Thus, increased oxygen from water quality improvements can increase the availability of metals and may cause metals that had been bound to sediments to leach into the water.

Hg can be readily taken up by worms living in the sediments. Polychaete worms are abundant and diverse in intertidal mudflats and are an important source for Hg biomagnification in food webs. Hg at the surface of the sediment is correlated with Hg in surface-feeding worms, but deeper burrowing worms contain greater Hg and meHg, showing that feeding ecology is important for predicting Hg bioaccumulation.

Mercury can be transported in the atmosphere far from its source—for example, to the Arctic. It appears that the long-range transport of mercury from Asia is an important source of atmospheric Hg to the Arctic. This Hg enters the water and becomes meHg, which is both toxic and biomagnifies in food webs. Mercury concentrations in organisms have increased and are controlled by a combination of meHg level, food web structure, and animal behavior (e.g., feeding behavior). Inuit people in the Arctic have high Hg in their blood and hair. Their reliance on traditional foods such as marine mammals, which are high in the food web, for subsistence means that they are particularly at risk from Hg exposure, even though they live very far from any Hg sources.

What are the toxic effects of different metals?

Mercury (Hg), especially meHg, is by far the most toxic. It affects embryonic development and is particularly neurotoxic, as is lead (Pb). Any chemical that affects the nervous system is likely to affect behavior at low concentrations. Reduced feeding and digestion are commonly observed after exposure to a variety of pollutants, including metals. Decreased feeding is not only a general response to contaminants, but the poor nutrition that results can in turn make animals more susceptible to contaminants and other stresses. Many organisms respond to reduced food intake with reduced activity, which may mean slower movements and reduced ability to catch food and to escape from predators. In this way, alterations in feeding and nutrient uptake can affect a population, and also could have ecosystem-wide repercussions.

Various metals reduce the respiration and metabolic rate of many organisms. Scientists usually measure oxygen consumption to determine changes in metabolic rates. Reduced respiration can be a response to reduced food intake as a way of conserving resources and energy. Metals (Cu, Pb, Zn, and Cd) also reduce the photosynthesis by the symbiotic algal cells (zooxanthellae) that live within coral animals and which are responsible for much of the nutrition of these animals. Exposure to metals, especially copper, can impair osmoregulation, the ability to maintain internal salt concentrations. Maintaining a constant internal chemical environment is particularly important in animals living in fluctuating environments such as estuaries. Some animals, called osmoregulators, maintain their body fluids at concentrations different from the surrounding water and must actively regulate salts. In the salty ocean, they drink seawater to offset water loss due to osmosis and then excrete the excess salt from the gills. This has an energy cost. If the animal moves to a lower salinity part of an estuary, it finds itself in an opposite environment—one where it will absorb water through osmosis, and where it must work to retain its salts and excrete this excess water. Exposure to contaminants can disrupt osmoregulation, which is performed primarily by enzymes in the gills.

Exposure to contaminants can alter reproduction. One particular concern is that very low levels of some environmental chemicals can interfere with the endocrine system, which is known as endocrine disruption. Metals and other contaminants can also directly affect gamete production, mating, and fertilization. These various stages of the reproductive process are clearly connected to one another. However, since most marine organisms normally produce enormous numbers of embryos, it is not clear how much reproductive impairment it takes to reduce the population size, which is rare.

One clear example of effects at the population level is that of TBT (tributyltin) on dog whelk snails. This chemical was formerly a very popular and effective component of antifouling paints used on boats. The first indication that there might be a problem with this effective antifoulant was observations in France by Alzieu and colleagues that oysters living near marinas looked abnormal. They grew very thick shells, became ball-shaped with very little meat inside, and were unmarketable. As a result of some excellent detective work, scientists traced the deformity back to the paint used on the boats that were concentrated in the marinas. A second observation was that dog whelk snail populations were crashing and female snails near marinas throughout Europe were growing penises. This condition was termed "imposex," and when severe, the male structures imposed on the female ones and blocked the oviduct so that eggs could not be deposited and females could not reproduce, as found by Peter Gibbs and Geoff Bryan. This effect was due to increased amounts of male hormones, which caused male reproductive structures to develop in the females. This abnormality, which rendered females sterile, led to drastic population declines. It also spurred the development of regulations, restrictions, and ultimately bans on the use of TBT on recreational boats. Imposex in female snails was discovered and studied several years before the term endocrine disruption was coined to describe such effects. Endocrine disruption will be discussed in greater detail in the chapter on organic contaminants (Chapter 6), since far more of them cause endocrine effects. Mollusks are not the only animals affected by TBT; they just are much more sensitive than others. They were affected at concentrations that were too low for the instruments in the late 1980s to measure, which spurred development of more sensitive instrumentation.

Early life stages are generally more susceptible to environmental contaminants than later stages. Embryos can be exposed to developmental toxicants such as meHg during egg development (oogenesis) in females, during the brief period between shedding of gametes into the water and fertilization, and after fertilization. Chemicals that are incorporated into the egg during oogenesis can produce malformations in the embryos that later develop from these eggs. Subsequently, embryos can be exposed to chemicals after fertilization. Exposures can be throughout embryonic development or during shorter time periods. Common responses include delayed development, formation of abnormalities, and reduced hatching. Some chemicals can produce effects that do not become apparent until later stages such as larvae or even adults. Most marine animals hatch out as small planktonic larvae with little resemblance to the adult form that they will eventually become. Larvae may be even more sensitive than embryonic stages of the same organism, since embryos are protected by an outer membrane that may reduce contaminant uptake and which is no longer present after hatching. Larvae also must usually swim and obtain food for themselves. Most benthic invertebrates have planktonic larvae, which at a certain stage of development must settle to the bottom to metamorphose into a juvenile in an appropriate habitat. Larval exposures to contaminants can lead to impaired settlement or to delayed physiological problems as juveniles or adults.

Developmental processes in later life can also be impaired by exposure to contaminants such as metals. Growth is an obvious and easily measured response that is frequently traced back to reduced food intake, but even without reduced feeding it may occur because organisms must expend energy to defend themselves against contaminants. The more energy needed to detoxify pollutants, the less will be available for growth. In addition to overall body growth, molting, regeneration, development of calcified structures (shell and bone), carcinogenesis (cancer), and smoltification (defined below) are other developmental processes that take place after larval stages and that are all sensitive to environmental contaminants such as metals. Salmon breed and embryos hatch upstream in shallow freshwater streams where they live for some time before migrating down to the sea. They undergo a developmental process called smoltification, which enables them to osmoregulate and live in salt water. This is hormonally controlled (by thyroid hormone) and can be impaired by a variety of contaminants. Aluminum (Al) is a normal constituent of soil and is generally not an environmental problem because it is not bioavailable. However, under low pH (acidic) conditions in freshwater that can result from acid rain, it becomes available. If a freshwater stream is affected by acid rain, young salmon get exposed to Al as well as to acidity, both of which cause toxic effects. Short-term Al-exposure and moderate acidification increase mortality in fish migrating downstream, and can reduce Atlantic salmon populations, according to Frode Kroglund and Bengt Finnstad. Al reduces their ability to adjust to salinity changes, so these young fish die when they approach seawater. Al also decreases growth hormone levels, and can cause stress and death.

Behavior is a particularly sensitive response to contaminants. Noticeable changes in behavior can be found at low concentrations of chemicals such as methylmercury (meHg), lead (Pb), or copper (Cu). In addition to being sensitive, behavioral changes can produce ecological effects at the population and community level, as shown by Weis and colleagues. Behaviors that have been studied include swimming, burrowing, migration, prey capture, predator avoidance, reproductive behaviors, aggression, and social interactions. Effects on behavior may be direct or indirect. Indirect effects include alteration of activity or reproductive success due to reduced feeding, and thus less energy. Behavior can be seen in larval or older stages after earlier exposure to pollutants during embryonic stages.

It is also possible to trace behavioral alterations back to underlying neurological effects of the contaminants. For example, Cu is especially toxic to the olfactory system of fish, which means that species that rely on their sense of smell for navigation or detecting the presence of predators (or prey) cannot do so as well. If salmon streams are contaminated with Cu, the very sensitive early stages of these migratory fish are damaged and their navigation during subsequent migration can be impaired, according to James Hansen and colleagues. Cu is also especially toxic to snails, which retreat into their shells and remain inactive while it is present. Oysters, on the other hand, can accumulate high concentrations of copper, so much that in high copper environments their tissues acquire a green coloration like the Statue of Liberty. Snails that eat these oysters get affected and become sluggish and reduce their feeding.

What can organisms do to defend themselves against metal toxicity?

There are a number of mechanisms that organisms use to reduce the toxicity of metals once they have taken them up. These include storing them in nontoxic forms such as metal-binding proteins or granules. Other mechanisms include stress proteins, and the evolution of tolerance. These are discussed in detail in Chapter 8.

Can elevated levels of metals in seafood be a risk to humans?

Since meHg biomagnifies in food webs, eating a lot of large fish that are high up in the food web, like tuna and swordfish, can be risky. Mean Hg concentrations for each type of seafood are highly variable. The high variability in Hg in common seafood has ramifications for public health and complicates the development of guidelines for how much should be eaten. Prenatal exposure to meHg has been associated with impaired performance on attention tasks, but the extent to which this translates into behavioral problems is not clear. In a study of Inuit mothers (who consume large quantities of fish) and their children, umbilical cord blood Hg concentration was associated with attention problems and ADHD symptoms in children, according to O. Boucher and colleagues.

Large tuna and swordfish together account for more than half of the Hg intake from seafood for the US population. American children may be ingesting high levels of Hg with their tunafish sandwiches. Research by the Mercury Policy Project, a consumer advocacy group (www.mercurypolicy.org) found that canned tuna has high levels of mercury. The group says albacore or white tuna can triple a child's Hg exposure, and recommends that schools and parents not serve it. The group also says children under 55 pounds should limit "light" tuna to one meal, once a month and twice a month for children over that weight. They also recommend that no child should eat tuna every day. The European Union recommends that pregnant or breastfeeding women not eat tuna more than twice a week, while the US Food and Drug Administration says they should avoid shark, swordfish, or king mackerel, but that some tuna should be included in their diet. Some feel these guidelines are out of date and stricter rules are needed to avoid the risk that even low levels of Hg could lead to problems in fetuses and young children.

What are the trends in metal pollution?

Overall decreases are seen, but improvement is very slow. Trace metals have been monitored in the Baltic Sea waters and biota with mixed trends. While the Baltic Sea remains much more contaminated than ocean waters, a slow decrease in dissolved Cd and Cu has been observed in many sites over a decade, but this was attributed to increasing hypoxia due to eutrophication, which precipitated the metals, rather than to an actual decrease in the metal contamination. Decreasing Pb was attributed to the switch to unleaded gasoline. Sediments, however, were still highly contaminated. Mercury and zinc also showed downward trends. Various fish species used for human consumption showed similar decreases. Declines in concentrations of Cd and Hg were seen in all the fish species studied by Lucyna Polak-Juszczak. Similarly, Pb in the livers of a number of fishes used for human consumption showed a significant declining trend.

What can be done to reduce metal pollution?

Metal pollution can be reduced by reducing inputs in the first place, and by remediating areas that have already been contaminated, which is the focus of the Superfund program of the

US Environmental Protection Agency. Superfund is the federal law that identifies the most highly contaminated sites in the country and eventually impels their cleanup. Technologies for remediating contaminated sediments are at various stages of development. Removing the contaminated sediments by dredging and subsequently putting them in a contained facility is the most common technique used. Volume reduction (i.e., removing only the sediments that require treatment and retaining as little water as possible) reduces costs; precision dredging techniques can reduce the volume of sediments that need remediation and thus reduce costs. Treatment costs may also be reduced through pretreatment. When sediments must be moved off-site for treatment or confinement, efficient hydraulic and mechanical methods are available for removal and transportation. Most dredging technologies can be used successfully to remove contaminated sediments.

What is natural attenuation?

Natural recovery (doing nothing and letting nature repair itself) is of low cost and, in some situations, may have the lowest risk of human and ecosystem exposure to contaminated sediments. It is most likely to be effective where surface sediment contaminant levels are relatively low and are being naturally covered over rapidly by cleaner sediments, or where natural processes destroy or modify the contaminants, so that contaminant releases to the water decrease over time. A disadvantage, however, is that the more contaminated sediments remain underneath, and could potentially be exposed if the overlying cleaner sediments are resuspended—for example, by severe storms.

What is capping?

Capping involves covering the contaminated sediments with a deep layer (typically one meter thick) of clean sediments, which forms the cap. Capping isolates the contaminated sediments and should prevent them from getting resuspended by storms. The original sediment bed must be able to support the cap, suitable capping materials (clean sand) must be available, and suitable water conditions (including depth) must exist to permit placement of the cap and to avoid compromising its integrity. Changes in the local substrate, burrowing benthic animals, or the depth at a site where new sediments are naturally deposited may subject the cap to erosion. A barrier to the use of capping is the Superfund legislation, which prefers more permanent controls, and capping is not considered to be a permanent control.

What is Confined Aquatic Disposal (CAD)?

Confined aquatic disposal (CAD) involves digging a deep pit in the bottom and placing the contaminated sediments inside it, then capping it. This technique is applicable to contaminated sites in shallow waters where capping is not possible and is good for the disposal and containment of slightly contaminated material from navigation dredging. Although the methodology has been developed, CAD has not been widely used. Among its advantages are that it can be performed with conventional dredging equipment and that the chemical environment surrounding the cap remains unchanged. Disadvantages include the possible loss of contaminated sediments during movement and placement.

What is bioremediation of metals?

Bioremediation is the process of using naturally occurring microbes to take up, digest, or convert waste material into harmless substances. Technologies can be generally classified as in situ or *ex situ*. In situ bioremediation involves treating the contaminated material at the site, while *ex situ* involves removing the contaminated material and treating it elsewhere. Bioremediation can occur naturally with the microorganisms already present at the site (natural attenuation or intrinsic bio-remediation), or can be accelerated with the addition of fertil-izers. Particular microbe strains can be added to enhance the resident microbe population's ability to remediate the contami-nants. Special genetically-engineered strains can be used. Not all contaminants, however, are easily treated by bioremediation. For example, metals such as Cd and Pb are not readily absorbed or captured by microbes. The assimilation of metals such as Hg into the food chain may have more negative than positive out-comes. Appropriate microbes should be tolerant to metals, have high metal-binding capacity, and synthesize metal-binding pro-teins in response to metal exposure. In some microbes, highly specific biochemical pathways have evolved to protect the microbial cells themselves from metal toxicity. A good example is the microbial reduction of mercury. In other cases, microbes can sequester toxic metals within soils or sediments or produce materials that can bind metals in contaminated soils. The mobi-lized metals can then be pumped out of the soil or sediment. Metals can be extracted from contaminated environments by two mechanisms. First, some heterotrophic microorganisms (those that need food, like animals) mobilize metals by produc-ing organic acids. Secondly, some highly specialized autotrophic bacteria (those that make their own food, like plants) generate large amounts of metal-leaching sulfuric acid from the oxida-tion of elemental sulfur. Some multiple metal tolerant fungi and bacteria have been identified that can be used to adsorb multi-ple metal pollutants. Bioremediation is cost-effective, and much less expensive than landfill disposal. New tools and techniques for use in bioremediation (e.g., genetically engineered organ-isms) are contributing to the rapid growth of this field.

What is phytoremediation?

Plants such as are found in salt marshes and mangroves can reduce the amount of metal pollution entering estuaries, and can remediate—to a degree—the contaminants in sediments.

Wetland sediments accumulate contaminants, and there are many cases in which wetland plants can remove pollutants, including metals. The use of plants to remove or stabilize contaminants is referred to as phytoremediation, and there are different mechanisms that can be utilized. Phytoremediation is a relatively recent technology and is perceived as cost-effective, efficient, and ecofriendly, with good public acceptance. It is an area of active current research. The approach in salt marshes is generally one of phytostabilization, where the plants are used to immobilize the metals and store them below ground in roots or soil, in contrast to phytoextraction, in which certain plants that can accumulate very high concentrations in aboveground tissues (hyperaccumulators) are used to remove metals from the soil and concentrate them above ground. Hyperaccumulators must in turn be harvested and disposed of to prevent recycling of the accumulated high concentrations of metals once they die and decompose. This is done frequently in terrestrial sites. However, wetland plants generally do not hyperaccumulate and, in any case, the mechanical aspects of harvesting plants would be destructive to wetlands with rooted plants.

Wetland sediments are generally considered a "sink" for metals and may contain very high concentrations of metals in a reduced state, especially in sediments with low oxygen. In such areas the bioavailability of the metals is very low compared to terrestrial systems, which have oxidized soils. Different forms of metals have different availability: water-soluble metals are the most available, while metals precipitated as inorganic compounds, or combined with humic materials are potentially available, and metals precipitated as insoluble sulfides (e.g., AVS) or bound within the structure of minerals are essentially unavailable. In estuaries, much of the metals are tightly bound to sulfides (e.g AVS) in anoxic sediments. However, some wetland plants (e.g., cordgrass, *Spartina alterniflora*) can oxidize the sediments near their roots by moving oxygen downward, and this oxidation can remobilize the metals, thus increasing their otherwise low availability. By oxidizing the soil right by the roots, plants can alter the distribution of metals in wetland sediments. Concentrations of metals tend to be higher in vegetated soils than in unvegetated ones, and are particularly high in soils near plant roots. Molecular tools are being used to better understand the mechanisms of metal uptake, translocation, sequestration, and tolerance in plants.

Salt marsh plants can absorb available metals from the sediments and store them largely in their roots. When wetland plants move metals up from root tissue they accumulate in leaves and stems. The degree of upward movement is dependent on the species of plant, the particular metal and various environmental conditions. Cordgrass transports significant levels of metals to aboveground tissues, and so plays a role in the transfer of metals through estuarine food webs. The metals that are moved up into stems and leaves become available to the marsh ecosystem if they are excreted from the leaves, which is what cordgrass does. Cordgrass has salt glands on the underside of its leaves for excreting salt; metals can be excreted from them as well. The release of metals from leaf tissues is a way for the plant to reduce its tissue levels of metals, but this increases the availability of metals in the ecosystem, with potential uptake into estuarine food webs. Metals not excreted from leaves will be in the leaves and stem when the plant dies, falls to the marsh surface and decays. The detritus produced as a result of this decay, will have elevated metals that will then be available to animals such as mollusks and small crustaceans that eat the detritus. In contrast to cordgrass, the invasive common reed *Phragmites australis* is widely used in constructed wetlands for treatment of wastewaters containing metals. *P. australis* concentrates more of its metals below ground in root and rhizome tissue, moving smaller amounts to aboveground tissues.

It has been shown that concentrations of metals such as iron, nickel, and chrome are 10 to 100 times higher than normal in mangroves downstream from mining sites and that mangrove trees absorb contaminants. However, their extent is decreasing in tropical areas because of human population growth and urbanization along coastlines, conversion of mangrove swamps to shrimp farms, and the use of their wood for fuel. Without the dense network of vegetation provided by the mangrove trees, sediments loaded with pollutants can enter lagoons and estuaries, which are sites of biodiversity and major sources of income for local populations through fishing and aquaculture.

6

PESTICIDES AND INDUSTRIAL ORGANIC CHEMICALS

What are the sources of pesticides to the marine environment?

Pesticides from agriculture, lawns, golf courses, and gardens wash into streams and rivers and ultimately down into estuaries. These chemicals are designed to kill agricultural pests (generally insects) on land. After being sprayed on land, they wash into the water when it rains and can affect aquatic animals. Some spraying happens directly in coastal habitats. Because salt marshes are well-known as breeding areas for mosquitoes, biting flies, and other nuisance insects, they are sites of direct pesticide applications. On the West Coast, where burrowing shrimp are considered pests in oyster-growing estuarine areas, the pesticide carbaryl is used to kill the shrimp. Estuarine and marsh organisms can also be exposed to herbicides used on the marshes to kill unwanted plants such as common reeds on the East Coast and cord grass on the West Coast. In addition to the pesticides used directly in salt marshes or estuaries, other insecticides and herbicides wash in from upland areas.

What happens to these chemicals after they enter the water?

Those chemicals of greatest concern are those that are persistent (i.e., that don't break down), that bioaccumulate in organisms, and that are toxic at very low concentrations. Some of

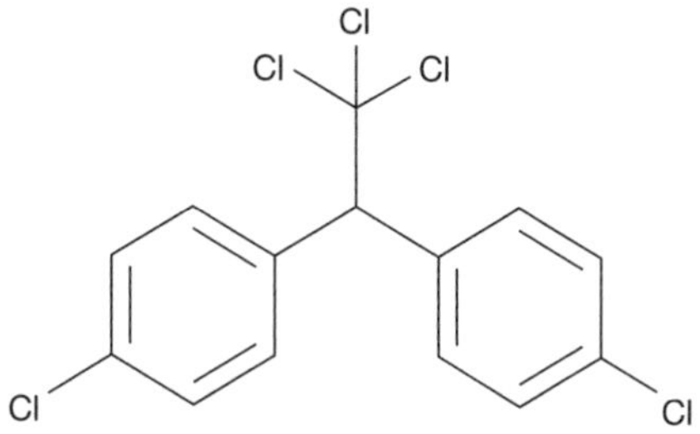

Figure 6.1 DDT molecule (in all organic molecules with hexagons, a carbon atom is at each point of the hexagon)

these, such as DDT, are banned in many countries, but they nevertheless persist in marine sediments. In some countries they are still used and continue to run off into aquatic environments. Organochlorine chemicals (mostly DDT-related pesticides, PCBs, and dioxins) have been studied intensively for decades. Dichlorodiphenyltrichloroethane (DDT) (Figure 6.1), the most powerful pesticide the world had ever known, can kill hundreds of different kinds of insects.

Its ability to kill insects was identified in 1939 by the chemist Paul Müller, who won the Nobel Prize for Physiology and Medicine. DDT was used in World War II to clear South Pacific islands of malaria-causing insects, and was used as a delousing powder. When it became available for civilian use it was considered a marvel, because it could be applied as a powder on the water in relatively small amounts and would keep killing mosquito larvae for months after only one application. It could kill all kinds of insects, was not particularly toxic to humans, and enjoyed great success until the development of resistance by both mosquitoes and eventually the public. DDT and related chemicals are fat-soluble and highly persistent. Insect populations can develop resistance because not all of the insects are killed when they are sprayed by the chemical. The few remaining resistant individuals breed, and their offspring are also more resistant to the chemical, an example of selection—evolution—at work. The insects eventually become so resistant that a different chemical has to be used. Because of their persistence in the environment and low water solubility, chlorinated hydrocarbons tend to accumulate in sediments and in tissues. Related pesticides included aldrin, dieldrin, chlordane, heptachlor, and toxaphene, which caused fish kills when applied near the water. Chlorinated hydrocarbons remain in the environment (especially sediments) for many decades, so they continue to be found long after they have been banned and continue to be sources of contamination to marine life. They can be moved by winds and currents far from their site of origin; for example, pollutants from Europe, Russia, and North America are transported to the Arctic. Furthermore, commercial bottom trawling (pulling fishing nets across the bottom to catch fish) churns up the sediments, releasing pollutants, as shown by Lycousis and Collins. Furthermore, animals can take up high levels of contaminants released by trawling. After only one month of exposure, mussels living near the bottom near trawling areas exceeded the EU limit for the chlorinated chemicals that cause developmental and reproductive abnormalities, so the high levels in edible mussels are of particular concern for public health.

Organisms can take up contaminants from the water, the sediments, and from their food, and may acquire tissue levels much greater than those in the environment. Not only do these chemicals remain in sediments for a very long time, they also biomagnify through food chains, increasing from one step to the next. DDT and other chlorinated hydrocarbons concentrate in fatty tissues. Animals accumulate and concentrate these chemicals from their food, and each trophic level will have greater concentrations than the level below it, so that the highest concentrations are in the top carnivores—big fish, predatory birds, marine mammals, and humans. Because of biomagnification, large carnivorous fish may have hazardous levels of contaminants, and health advisories may be issued to protect humans from consuming them.

What is the importance of the book Silent Spring?

Rachel Carson's writing about the dangers of DDT was stimulated by bird kills that she observed following DDT sprayings. *Silent Spring* described how DDT entered the food chain and accumulated in fatty tissues of animals, including humans, causing cancer and genetic damage. She noted that a single application on a crop killed insects for weeks and months (not only the targeted insects but many others), and remained toxic even after dilution by rainwater. She concluded that DDT and other chlorinated pesticides had harmed birds and other animals and had contaminated the world food supply. The book alarmed readers and triggered an indignant response from chemical industry spokesmen, who said that if people were to follow her advice we would return to the Dark Ages, and insects and diseases would inherit the earth. Anticipating such a reaction, Carson had written the book with numerous scientific citations and a list of scientific experts who had approved it. Many eminent scientists, as well as President Kennedy's Science Advisory Committee, supported the book. As a result, DDT came under much closer government scrutiny and was eventually banned. Most of the other chlorinated hydrocarbons were also gradually phased out in subsequent decades. An important legacy of *Silent Spring* was a new public awareness that nature was vulnerable to human activities. The growth of the environmental movement was partly a response to this new awareness. Most uses of DDT and other chlorinated hydrocarbons were banned in the 1970s. In the United States, the Federal Insecticide, Fungicide, and Rodenticide Act (FIFRA) requires that adverse ecological effects be balanced against the economic costs of regulating pesticide use and the benefits the pesticide provides.

What are some newer types of pesticides?

Since the 1960s, the variety of pesticides has increased greatly. Hundreds of chemicals are now in use, and they generally occur in mixtures. Newer chemicals are less persistent than the chlorinated hydrocarbon pesticides and do not generally cause fish kills. However, they can produce sublethal effects such as endocrine disruption, altered development and behavior, reduced growth, and other effects. "Second-generation" pesticides such as organophosphates and carbamates are much less persistent in the environment, but if spraying coincides with the time of reproduction and early life stages of susceptible organisms, they can harm sensitive early life stages. Organophosphates break down in the environment in a matter of weeks. They were developed from chemical compounds similar to nerve gas and, not surprisingly, they affect a chemical in the body that is important for the transmission of nerve impulses. At high doses, organophosphates can overstimulate the nervous system and cause nausea, dizziness, or in cases of severe poisoning, convulsions and respiratory paralysis. One organophosphate commonly used in salt marshes is temephos (Abate®), which is considered hazardous to fish, birds, insects (beneficial species as well as the pests), shrimp, and crabs. Reductions in fiddler crabs and zooplankton have been seen after its use for mosquito control, and it was found to accumulate in salt marsh organisms, including sheepshead minnows, mussels, and fiddler crabs. Malathion can be applied as a fog from a moving vehicle, and it will permeate through vegetation, killing adult mosquitoes. It is considered one of the safest organophosphates, and has been used in large pest eradication programs. However, honeybees are quite sensitive to it, and colonies are sometimes affected downwind from an application. It degrades rapidly in the environment, especially in moist soil, and has relatively low toxicity to estuarine organisms, birds, and mammals. It is usually broken down within a few weeks by water and sunlight and by bacteria in soil and water, but it can affect nontarget estuarine organisms before it is completely broken down. Current use pesticides are frequently detected in the environment and tissues of animals, even though they are less persistent than the earlier chemicals.

Many different pesticides (fungicides, herbicides, and insecticides) are found in water, sediments, and aquatic organisms of estuaries in agricultural areas.

What are "third-generation" pesticides?

As people have realized the widespread effects of both first- and second-generation pesticides, attempts have been made to develop new pesticides that are more specialized in their toxicity to insects. Insect growth regulators are much less toxic to birds, mammals, and fishes. Some of these newer pesticides target the molt cycle of insect larvae by mimicking their specific biology or hormones. Larvicides target the insect's larval stages and are less harmful to nontarget organisms, and generally more effective and specific than chemicals that focus on adults. Larvicides target the limited breeding habitat before adults have had a chance to disperse widely. One larvicide in use is methoprene, an insect growth regulator that mimics the insect's juvenile hormone (JH), which normally prevents larvae from metamorphosing prematurely into adults. When the insect stops secreting JH during the pupal stage, it is then ready to develop into an adult. If methoprene is present in the insect's system when it begins the pupal stage, the triggering of adult development is prevented and it dies as a pupa. While few impacts have been observed in nontarget aquatic organisms, there is concern that these chemicals might have harmful effects on crustaceans, which are closely related to insects. Another larvicide is Dimilin® (diflubenzuron), a chitin synthesis inhibitor. Chitin is a major constituent of the outer exoskeleton of arthropods (including both target insects and nontarget crustaceans). A chitin synthesis inhibitor prevents the larvae from molting, resulting in their death. Unfortunately for crustaceans, they have the same chitin in their exoskeletons and also need to molt, so they can be severely harmed when diflubenzuron is sprayed. Crustaceans make up the majority of small animals in the zooplankton, and reduced numbers of zooplankton means less food for small fishes. Also affected are larger crustaceans such as shrimps and crabs, particularly when they are in larval stages and must molt frequently.

Bacteria that cause insect diseases can also be used as pesticides. The principal one used on salt marshes is *Bacillus thuringiensis israelensis* (Bti), which produces protein crystals that are selectively toxic to mosquito larvae. After being eaten, they rupture the digestive tract of the host, causing rapid death. When specific diseases of insects are used as pesticides there is less likelihood of harm to nontarget organisms, but honeybees, butterflies, dragonflies, and other useful insects may also be at risk.

How are pesticides regulated?

In the United States, FIFRA requires that the adverse ecological outcomes of a pesticide be balanced against its beneficial effects in controlling pest populations (e.g., increased agricultural production). In registering pesticides for use, it has always been easier to document its financial benefits than to estimate ecological costs, and the toxicity testing is done by the manufacturer of the pesticide, who has an interest in minimizing its adverse effects. Most of the toxicity testing under FIFRA is based on lethal effects (LC_{50}, or the concentration that kills 50% of the exposed animals) rather than sublethal effects. The LC_{50} is not relevant to real-world effects of pesticides, since it does not encompass species differences, sublethal effects, or delayed effects. Most of the standard aquatic test organisms are freshwater species—rainbow trout, bluegill, and daphnids. Individual chemicals are evaluated by these standardized tests; in nature, however, animals are exposed to a variety of chemicals, with new ones coming into use every year, and the toxicity of complex mixtures is not well understood at all.

What is integrated pest management?

Insect control is slowly evolving from a reliance on chemical insecticides to "integrated pest management" that includes surveillance of pest populations, source reduction, larvicides, and biological control. Surveillance programs in salt marshes track adults, larvae, and larval habitats, and only when pest populations exceed some set level are control activities initiated. Source reduction involves elimination of larval habitats, and includes open marsh water management and rotational impoundment management where the marsh is minimally flooded during the summer by temporary impoundments— reducing mosquito breeding. Biological control includes the use of predators to eat mosquito larvae, such as aquatic invertebrates, mosquitofish, and killifish.

What are the effects of pesticides on nontarget organisms?

In the 1960s, predatory birds such as brown pelicans, eagles, and ospreys in coastal areas of the United States accumulated such high levels of DDT and related pesticides that they had reproductive failure and were in danger of becoming extinct until the chemical was banned in 1970. DDT and related pesticides caused these birds to lay eggs with very thin eggshells, so that the eggs broke when the birds sat on them, resulting in reproductive failure.

Toxic effects can be lethal or sublethal. Effects can be documented in laboratory exposures or observed in organisms in the field. Effects can be studied at various levels of biological organization from the molecular level (e.g., effects on enzymes or DNA), to the organism (e.g., effects on growth, physiology, behavior, development), to the population (e.g., changes in population density, birth rate, or age structure), to the community (e.g., effects on the number of species present). Effects on the organisms' immune system, endocrine system, nervous system, reproductive system, and so on are critical. Of great concern are chemicals with the ability to cause embryonic malformations (teratogenesis), genetic alterations (mutagenesis), endocrine disruption, or cancer (carcinogenesis).

Chlorinated hydrocarbons such as DDT, which biomagnify and take a long time to break down, can prevent marine larvae from developing normally, reduce respiration and metabolism, impair growth, and impair salt and water balance. Organophosphate and carbamate pesticides, the "second-generation" pesticides, still have harmful effects. The insecticide Sevin® (carbaryl), used to control ghost shrimp in Pacific oyster beds, is also very toxic to commercially important Dungeness crabs. Their larvae are highly sensitive to other insecticides and fungicides as well, as reviewed by Feldman and colleagues. Malathion slows down larval development by delaying molting; Abate® affects behavior, making animals more susceptible to predators.

What is endocrine disruption?

There is particular concern about chemicals that, even at very low concentrations, alter the development and functioning of the endocrine system and affect development. These chemicals are called endocrine disruptors, a term coined by Theo Colborn. Known endocrine disruptors include DDT and other chlorinated hydrocarbon pesticides, certain PCBs and dioxins, some metals, plastics, detergents, and flame retardants. These chemicals have different mechanisms of action, depending on the life stage at which the animal is exposed, and they may have effects that are not seen for years after exposure. The most commonly studied chemicals are those affecting reproduction, and they may mimic natural hormones or inhibit them so that reproduction may be disrupted, intersex offspring may be produced, and metamorphosis may be delayed, accelerated, or prevented.

The first documented examples of endocrine disruption in the estuarine environment were in dog whelks and mud snails that were affected by the use of tributyltin (TBT) in antifouling paints applied to boats to prevent algae and barnacles from attaching to the hull (see Chapter 5). Previously, copper had been the main toxic ingredient in these paints, but TBT was more effective and long-lasting. As discussed in the previous chapter, tin-based chemicals, even at extremely low levels, caused female snails to develop male sexual organs and to become sterile, and populations crashed. TBT is now prohibited in marine paints in most countries, and snail populations are recovering. Most boats are again being painted with copper-based paints that are far from benign but are much less toxic than TBT, and research efforts are underway to find nontoxic or less toxic methods to prevent fouling. One popular formulation being used is adding a substance called irgarol to copper-based paints to boost their effectiveness. Though not an endocrine disruptor, irgarol is highly toxic to plants, including phytoplankton, seaweeds, and seagrasses. It is fairly stable in water and sediment, and has become a widespread contaminant in the vicinity of marinas and poses a continual risk to the environment.

Some other pesticides and industrial chemicals in very low concentrations also may affect hormone functions, and it is suspected that reported decreases in human sperm counts and increases in sperm abnormalities may be a result of exposure. Many reproductive abnormalities in different animals have been reported in nature. Alligators from Lake Apopka in Florida that were exposed to pesticides had reduced penis size and sperm abnormalities. Intersex frogs appear in areas where the herbicide atrazine is used. Mosquitofish near paper mills have intersex conditions in which females grow an extended anal fin called a gonopodium, typically seen only in males. The eggshell thinning noted in birds exposed to DDT was probably also an example of endocrine disruption, although that term had not yet been coined when the problem occurred. Endocrine disrupting chemicals appear to be involved with increased incidence of hermaphroditism in some fishes and marine mammals. Polar bears, living far away from any source of contaminants but at the highest trophic level in the Arctic where contaminants concentrate, seem to be exhibiting abnormal genitalia.

What are biomarkers?

Biomarkers are changes (generally biochemical) that can be used to assess responses to contaminants. Some molecular biomarkers include the induction of cytochrome P-450 1A (or CYP1A), which indicates exposure to aromatic hydrocarbons and is an enzyme system that is used to metabolize them. Another biomarker is vitellogenin (yolk protein) production in males, which indicates exposure to estrogenic chemicals and is normally produced only by females. The eggshell thinning of birds described above could also be considered a biomarker.

What kinds of population level effects can be produced?

Population level effects such as reduced numbers can be seen from chemicals that do not necessarily have direct impacts on reproduction. For example, chemicals that are neurotoxic can affect behavior, including reproductive behavior. If reproductive behavior is abnormal, reproduction will be impaired and the population may diminish. If the chemicals affect feeding behavior and the organisms eat less, they will not grow as well and may not be able to reproduce as well or live as long as unaffected individuals. If contaminants impair the ability to avoid predators, the animals will not live as long and may not leave offspring. If chemicals affect the immune system, the organisms will be more vulnerable to diseases. These are just some of the ways that populations could decrease as a result of contaminant exposure. Populations in chronically polluted areas can also respond by becoming more tolerant to the contaminants, which can select for more tolerant individuals. This selection (an evolutionary response) results in a genetically different population.

What community level effects can be produced?

Once populations of sensitive species are affected, changes in communities can result. In general, communities become less diverse because of the loss of some sensitive species. Shifts in community composition also occur in which tolerant species become more abundant, while the more sensitive species decline. Community-level effects are most commonly studied in benthic communities that are composed largely of polychaete worms and bivalve mollusks that live in the sediments and cannot move away quickly. A useful approach devised by Peter Chapman is to measure concentrations of contaminants in the sediments, the toxicity of the sediments (by sediment toxicity tests), and the benthic community structure. This is referred to as the sediment quality triad. While it cannot indicate which particular contaminants are responsible for toxicity, it is very useful for comparing different areas or changes in a given area over time. Contaminated sites tend to have multiple contaminants that may interact in different ways. Because of these interactions, it is very difficult to predict biological effects based only on knowledge of the types and concentrations of contaminants at a particular site.

What can marine organisms to do defend themselves against toxic effects?

Animals have enzyme systems that can detoxify organic chemicals and break them down. These are discussed in Chapter 8. Over the long term, chemicals can select for individuals that are more tolerant, and thus evolution of more tolerant populations may take place.

What are the trends in pesticide contamination?

Persistent contaminants in coastal sediments and biota in the United States have been generally decreasing over the past two decades. The National Oceanic and Atmospheric Administration (NOAA) has run a monitoring program for 140 contaminants in bivalve populations (mussels and oysters) in 300 sites nationwide. For butyltins (TBT having been banned), 88 sites showed a significant decrease while none showed an increase. For organic contaminants such as chlorinated hydrocarbon pesticides (most of which were banned in the United States in the 1970s and 1980s), 133 sites showed a significant decrease while none showed an increase. The Canadian government has been monitoring contaminants in bird eggs and has found decreases in chlordane, dieldrin, and DDT-breakdown product DDE (banned pesticides) in eggs of the great blue heron, double-crested cormorant, and osprey.

What are polychlorinated biphenyls (PCBs)?

Polychlorinated biphenyls (PCBs) are also chlorinated hydrocarbons, and were manufactured from 1929 until they were banned in the United States in 1979 (Figure 6.2). PCBs are chemically related to organochlorine pesticides. Each molecule consists of chlorine atoms attached to a double carbon-hydrogen ring (a biphenyl). There are 209 different PCB molecules (or congeners) that differ in the number and location of the chlorine atoms on the rings. In general, PCBs with more chlorine atoms are more toxic than PCBs with less chlorine. Like chlorinated hydrocarbon pesticides, they remain in the environment for a long time, have low water solubility, and accumulate in fat. PCBs are suspected of causing cancer and have been linked to male sterility and birth defects. In birds and fish they decrease egg hatchability, alter behavior, and decrease immune response. There are two distinct categories of PCBs: coplanar and non-coplanar (or *ortho*-substituted) congeners. Coplanar

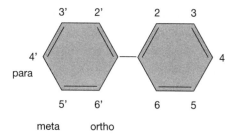

Figure 6.2 PCB (polychlorinated biphenyl) molecule, showing the various sites where Cl atoms may be attached to carbon

PCBs have a fairly rigid flat structure, with the two phenyl rings in the same plane, which gives the molecule a structure similar to polychlorinated dibenzo-p-dioxins (PCDDs; see below) and allows them to act in the same way as these molecules. Non-coplanar PCBs, with chlorine atoms at the *ortho* positions, are not part of the extremely toxic dioxin group. Nevertheless they have neurotoxic and immunotoxic effects, but not at such low levels as those related to dioxins. Due to their nonflammability, chemical stability, high boiling point, and electrical insulating properties, PCBs had hundreds of industrial and commercial applications including electrical equipment; as plasticizers in paints and rubber products; in pigments, dyes, and copy paper; and many others.

How did PCBs get into the marine environment?

Unlike pesticides, PCBs were never intentionally sprayed in the environment. They were used in industry and got into the environment through carelessness during their manufacture and use. They can still be released into the environment from poorly maintained hazardous waste sites, illegal or improper dumping of PCB wastes, leaks from electrical transformers containing PCBs, and disposal of PCB-containing products into landfills not designed for hazardous waste. PCBs may also be released to the environment by the burning of some wastes. They can be carried long distances and have been found in snow and seawater in areas far away from where they were released, and therefore they are found all over the world. In general, the lighter the PCB (the fewer chlorine atoms per molecule), the further it can be transported. PCBs accumulate in aquatic biota, including plankton and fish, and like chlorinated pesticides and meHg, PCBs biomagnify. Thus, fishes higher on the food chain have higher concentrations than smaller ones (Figure 1.3).

General Electric plants in upstate New York dumped an estimated 1.3 million pounds of PCBs into the upper Hudson River over a 30-year period until they were ordered to stop in 1977. Since that time, the spread of PCBs throughout the river created a widespread toxic waste problem. The contaminated sediments have dispersed to cover a much larger stretch of the river than they did originally, making the cleanup more extensive and far more expensive. An approximately 200-mile stretch of the river is designated a Superfund site. Though required by the law to clean up the PCBs, General Electric battled the EPA in the courts for decades until an agreement was finally reached in 2005 and cleanup finally started in 2011.

The New York State Department of Environmental Conservation initiated intensive monitoring of PCBs in fish and banned commercial fishing for striped bass in the region in 1976 following the discovery of high levels in this species. Since then, the PCBs levels in the fish around New York Harbor have dropped and then stabilized at an acceptable level. PCBs in fish in the upper Hudson, however, still exceed what is considered safe, according to a 2013 report by the Hudson River Natural Resource Trustees (comprised of the State of New York, Department of Commerce—NOAA, and Department of Interior). Most of the PCBs in the lower Hudson originate from the upper Hudson, but about 40% of the elevated levels in the New York/New Jersey Harbor come from local sources.

PCB levels in white perch from the Chesapeake Bay area were found by King and colleagues to be strongly related to the percentage of suburban and urban development in the local watershed. They considered the intensity of development in watersheds using four measures of developed land-use (% impervious surface, % total developed land, % high-intensity residential + commercial, and % commercial) to represent potential source areas of PCBs to the subestuaries. When development of the land in the watershed reached about 20% of the total area (which is not particularly dense for our coastal states), PCBs in the fish begin to exceed recommended limits for consumption. PCBs historically produced or used in commercial and residential areas are apparently persisting in the environment and the amount of developed land close to the subestuary had the greatest effect on PCB levels in the fish.

PCBs have been banned in the United States since the 1970s, but continue to be redistributed and dispersed. More than 30 years after their prohibition, they are still accumulating in fish tissue to such an extent that state agencies recommend that people do not eat striped bass or blue crabs from the Newark Bay area, and eat no more than one meal a week of seafood from other areas in the New York Harbor estuary.

Another PCB-contaminated site is an 18,000 acre tidal estuary in New Bedford, Massachusetts, where manufacturers of electric devices used PCBs and discharged wastes into the harbor directly as well as through the city's sewage system. PCB levels in fish and lobsters exceed the Food and Drug Administration's limit for PCBs in edible seafood. There is an increased risk of cancer and other diseases for people who regularly eat seafood from the area. While some species have disappeared, Diane Nacci and colleagues at EPA found that the killifish at the site have become very tolerant to the PCBs. The fish adapted to the high levels of PCBs through genetic changes by developing an abnormal biochemical pathway.

What are Dioxins?

Dioxins and furans are some of the most toxic chemicals known. Dioxin is a general term that describes a group of hundreds of chemicals that are highly persistent in the environment. The most toxic compound is 2,3,7,8-tetrachlorodibenzo-p-dioxin or TCDD (Figure 6.3). Polychlorinated dibenzofurans are similar to dibenzodioxins, but with a single oxygen connecting the two benzene rings instead of two oxygens.

The toxicity of other dioxins and dioxin-like PCBs are measured in relation to TCDD. Dioxins and furans are not made on purpose, but are formed as unintentional byproducts of industrial processes that use chlorine, such as chemical and pesticide manufacturing, pulp and paper mills that use chlorine bleach, the production of polyvinyl chloride (PVC) plastics, the production of chlorinated chemicals, and incineration of wastes containing plastic. Dioxin, a contaminant in the herbicide Agent Orange, was found at Love Canal in Niagara Falls, New York, and was the basis for evacuations at Times Beach, Missouri and Seveso, Italy. The industrial accident in Seveso led to many cases of Acquired Dioxin-Induced Skin Haematoma, in which the skin acquires disfiguring red lumps.

Dioxins, like PCBs, are organic molecules with varying numbers and arrangements of chlorine atoms. They are particularly toxic to the immune system and to developing embryos, in which effects may occur immediately or may be delayed for a long period of time, perhaps impairing reproduction once

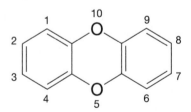

Figure 6.3 Dioxin molecule, showing the various positions where Cl atoms may be attached. 2,3,7,8 TCDD is the most toxic

the individual matures. They are known to alter hormones, and to cause reproductive problems, liver damage, wasting syndrome, and cancer. Judging from the Seveso and other incidents, humans appear to be much less susceptible to the immediate acute effects than other species. At Seveso, most farm animals died, while people just got skin hematoma.

How do they get into the marine environment?

There is an estuarine/marsh dioxin-contaminated site in Newark in the lower Passaic River right before it enters Newark Bay. The Diamond Shamrock Company manufactured the herbicide Agent Orange there during the Vietnam War and released dioxins as a byproduct. Newark Bay and the lower Passaic have layers of polluted sediment contributing to dangerous dioxin levels in blue crabs, fish, and fish-eating birds. The dioxin levels recorded in Passaic River and Newark Bay blue crabs are among the highest ever measured in aquatic animals, and crabbing is banned in the area because the contamination poses a high cancer risk.

The presence of dioxins in Newark Bay sediments has made the disposal of dredged materials from deepening the channel for the Port of Newark a highly controversial issue. In February 2004, the EPA determined that pollution in the bay posed an imminent and substantial risk to human health and to the environment, and it ordered the company that was responsible for the pollution in the lower Passaic River to carry out a comprehensive study under EPA supervision and to design a cleanup plan. At this site and at many others in need of remediation, there is concern that when the bottom is dredged as part of the cleanup process, sediments could be stirred up and contaminants that are tightly bound to the sediments will be mobilized and released, causing additional exposure and risk of toxicity to animals and plants in the area. Under the Superfund Program, the clean up of the lower Passaic finally started in 2012.

What effects do they have?

Dioxins, like chlorinated hydrocarbons, are persistent in sediments of marine systems, where their effects continue long after they are banned. They also biomagnify through food webs, and long-term effects are generally not well known. Fish embryos are highly sensitive and develop a syndrome that prevents their normal development. Ironically, pollution can have some indirect positive effects on crabs. Blue crabs in industrialized northern New Jersey accumulate such high concentrations of dioxin that fishing for them is banned out of concern for human safety. Consequently, their population is growing and individuals grow larger than in clean areas where they are still fished for.

Can PCBs or dioxins be a risk to people who eat seafood?

People who ingest fish or other seafood with high levels of PCBs or dioxins, such as in Newark Bay may be exposed to high concentrations. Toxic fish can be found in many other urbanized estuaries. Fish from the Columbia River near Portland, Oregon have PCB levels thousands of times above what the EPA considers safe for unrestricted consumption. Many rivers in the Columbia Basin, including sections of the mainstem Columbia River, have fish advisories that warn people not to consume certain types of fish, but people do not always heed the signs.

Maternal exposure to PCBs and dioxins was associated with an increased risk of asthma and more frequent upper respiratory tract infections in babies. Furthermore, maternal exposure to PCBs and dioxins was also associated with reduced antibody response to a measles vaccine. Thus, prenatal dietary exposure to PCBs and dioxins may increase the risk of asthma and susceptibility to infectious diseases in early childhood.

What can be done about organic chemical pollution?

Obviously, removing or cleaning up contaminated sediments will result in a cleaner environment over time. Fortunately, PCB and dioxin levels have been declining in the past few decades and have been the subject of a number of federal and state regulations and cleanup actions in the United States. A number of Superfund projects have removed highly contaminated sediments (e.g., Hudson River PCBs). One looks forward to the day, probably decades from now, when people can safely consume fish and crabs from around Newark Bay and the rest of New York/New Jersey Harbor, New Bedford Harbor, the Columbia River estuary, and other contaminated sites.

7

EMERGING CONCERNS

What other types of contaminants are we beginning to learn about?

Contaminants of emerging concern (CECs) have been defined as any man-made or naturally occurring chemical or microorganism that is not generally monitored in the environment but has the potential to enter the environment and cause adverse ecological or human health effects. In some cases release of these contaminants has been going on for a long time, but they have not previously been regarded as contaminants and are already widespread. They can come from municipal, agricultural, and industrial wastewater sources. Some examples are: brominated flame retardants, phthalates (plasticizers), alkylphenols (used as detergents and known to disrupt the reproductive system), pharmaceuticals, and triclosan (trichlorohydroxydiphenyl ether), an antibacterial agent in many personal care products and which poses risks to humans and the environment. There are also a plethora of new chemicals that have recently come into use (e.g., some drugs, nanoparticles). They came into widespread use before we know anything about their environmental impacts.

Why are pharmaceuticals and personal care products (PPCPs) a concern?

Pharmaceuticals are prescription and over-the-counter drugs, including antibiotics, birth control pills, tranquilizers, painkillers, and other medications, while personal care products include soaps, fragrances, sunscreen, and cosmetics. Even caffeine (in food and beverages as well as some pharmaceuticals) has been found in coastal waters. There are mounting concerns about pharmaceuticals that are being found wherever they have been looked for in waterbodies. There are several reasons for the concern. Large quantities of pharmaceuticals and personal care products (PPCPs) enter the environment after use, and sewage systems are not equipped to remove them. Most treatment plants filter and chlorinate sewage to remove disease-causing microbes and excess organic matter but do not remove pharmaceuticals, which go right through traditional treatment processes. When treatment plants release treated sewage (effluent), drug-tainted water is released directly into the receiving waterbody. Researchers such as Daughton and Kearns have found antibiotics, blood-pressure reducers, hormones, psychiatric drugs, and painkillers in the water leaving sewage plants and in the waterbodies receiving this wastewater. The risks posed to aquatic organisms and to humans are unknown, because the concentrations are so low. Since pharmaceuticals are designed to have biological effects at very low concentrations, it is not surprising that they should have effects on aquatic organisms. Two of the major concerns about pharmaceutical pollution have been the development of resistance to antibiotics by microbes and endocrine disruption by natural and synthetic sex steroids (such as birth control pills). Many other PPCPs have unknown effects. These contaminants are being discovered in water and fish tissue at very low concentrations. It is likely they have been present in the environment for as long as they have been in use. Many PPCPs remain in the water because as they degrade more are continually being added, and their use is increasing. Because of increasing concentrations, environmental effects are being noticed.

When endocrine disruption was first being studied in aquatic animals in the early 1990s, people looked to the "usual suspects," the chlorinated hydrocarbons (which do have endocrine effects). Later it was noticed that actual hormones themselves were in the water, coming out from sewage treatment plants, and scientists realized that estrogens from birth control pills were playing a major role. Waters contain pharmaceuticals, such as metabolized birth control pills, that people excrete and which can affect fish and other animals. Currently, there is considerable interest in investigating the pharmaceuticals and personal care products that end up in aquatic ecosystems, and numerous studies have found endocrine effects in aquatic organisms. Altered sex ratios and abnormal female fish have been seen downstream of treatment plants, as well as intersex fish with both male and female reproductive tissue. It would not be at all surprising if pharmaceutical pollution produces greater endocrine effects than the usual suspect pollutant chemicals, because the biology of humans is very similar to that of fish in this regard. When impacts of CECs from wastewater were investigated in Southern California by Steven Bay from the Southern California Coastal Water Research Project and colleagues, CECs were found in effluents from the major municipal wastewater dischargers, as well as in seawater, sediments, and fish near the outfalls. Fish hormones were altered; they had reduced stress response, altered estrogen, and reduced thyroid hormone, but responses could not be definitively linked to the discharges. However, thyroxine was lower in fish from all discharge sites, and estradiol was lower at three of the four outfall sites. The physiological changes, however, did not apparently lead to decreased reproduction.

Glucocorticoids (adrenal hormones) are also found in the environment at concentrations that may be high enough to affect aquatic life. A team of scientists that reviewed hundreds of studies concluded that there are no safe doses for hormone-altering chemicals. Such chemicals have effects at low levels, which are often completely different than effects at high concentrations.

Tranquilizers, antidepressants, and other neuroactive pharmaceuticals may affect the behavior of fish and wildlife. Beulig and Fowler studied "fish on prozac"—the effects of the selective serotonin reuptake inhibitor (SSRI) fluoxetine (Prozac) on fish. They found that it alters the amount of the neurotransmitter serotonin (just as it is designed to do in humans), which causes the fish to reduce their swimming and feeding behavior. It also has some toxic effects on algae. Even low levels of oxazepam caused perch to become more antisocial, risk-prone, and hyperactive, making them easier targets for predators. Only about 20% of the dose of commonly used cholesterol lowering drugs (statins) is metabolized in the body; the rest is excreted and finds its way into aquatic systems, where their effects are largely unknown. Antibiotics such as erythromycin and tetracycline can stimulate the evolution of (or selection for) antibiotic-resistant bacteria that can later cause illness in wildlife and humans. Also detected are caffeine, nicotine, acetaminophen, ibuprofen, and many other familiar substances—a veritable drug cocktail. While they are not likely to be toxic in the traditional sense, they are biologically active and likely to have effects on aquatic animals.

In addition to pharmaceuticals, personal care products such as cosmetics, lotions, sun blocks, and insect repellants (for example, DEET) are not broken down or removed in sewage treatment plants, and when they enter aquatic systems the chemicals in these products can affect aquatic plants and animals. Triclosan, an antibacterial commonly used in personal care and household products, is one of the most frequently found chemicals in wastewater in the United States. It is a potent endocrine disruptor with effects on the thyroid gland, and is also toxic to aquatic plants. Furthermore, there is no evidence that over-the-counter antibacterial soap products are any more effective at preventing illness than washing with plain soap and water. Other antibacterial compounds are released from sewage treatment plants, and wherever they have been looked for, they have been found.

What can be done about PPCPs?

Education is a major approach to pharmaceutical pollution. "Don't Rush to Flush" became the motto used to teach the public about the risks of flushing or improperly disposing of unwanted or unused over-the-counter medications, pharmaceuticals, and personal care products. However, most of the problem is not due to unused medications improperly flushed down the toilet, but those that were taken properly and later excreted in urine. Conventional methods of filtering wastewater in sewage treatment plants can't completely remove medicine residues. New technologies for treatment plants are needed, but not yet ready for use. Chemical processes are being devised and tested that can remove persistent chemicals and pharmaceuticals from wastewater. Pollutants can also be removed effectively from wastewater with selective adsorbers. A biological filter has been developed in which specific enzymes (called laccases) break down pharmaceuticals. These are still in the experimental stage and have not been put to use anywhere, but provide an idea as to approaches that can be taken. Another new water treatment technology called membrane distillation separates drug residues from sewage by heating. Water vapor passes through a thin membrane and through an air gap, where it condenses onto a cold surface. Drug residues collect on one side of the membrane and water on the other. In a test with oxazepam in wastewater, the level was reduced to less than 1% of the original concentration. This technology is also in the very early stages of development.

What are polybrominated diphenyl ethers (PBDEs) and why are they a concern in the marine environment?

Other emerging contaminants are flame retardants that are used in a variety of consumer products including clothing, furniture, curtains, carpets, and toys. They are intended to slow the rate of ignition and fire spread, giving people time to escape from a fire or extinguish it. They have been found at very high levels in aquatic systems and are also common in landfills. Attention to these contaminants is recent, since no one had bothered to look for them before. The chemicals are polybrominated diphenyl ethers (PBDEs); their structure is reminiscent of PCBs and dioxins, but with bromine attached instead of chlorine, and (not surprisingly, given the similar chemistry of chlorine and bromine) they are also persistent, toxic, and bioaccumulative (Figure 7.1). They also have many neurological, endocrine, and developmental effects similar to PCBs and dioxins, and they are extremely potent thyroid hormone disruptors. In recent years PBDEs have generated international concern over their widespread distribution in the environment, their potential to bioaccumulate in humans and wildlife, and their suspected adverse health effects. Production of PBDEs in the United States began in the 1970s and peaked in the late 1990s. An investigative report from the *Chicago Tribune* suggests that their widespread use was pushed by the tobacco companies, which were under fire (as it were) for cigarettes causing house fires. The industry insisted it couldn't make a fire-safe cigarette that would appeal to smokers and instead promoted flame retardant chemicals in furniture—shifting attention from cigarettes to the couches and chairs that were going up in flames. They pushed the use of retardant chemicals in furniture and even got the fire marshals association to promote it—even after it was found that the chemicals were escaping and accumulating in people and the environment. With furniture treated with flame retardants, people could still smoke but not die in burning houses. That way they could continue buying cigarettes and smoke for more years (until they died of lung cancer).

PBDEs move from consumer products to the outdoor environment and have been found by Barry Kelly and colleagues in tissues of marine mammals in the Arctic, far from any consumer products. PBDE concentrations in the US marine environment are among the highest in the world, perhaps because

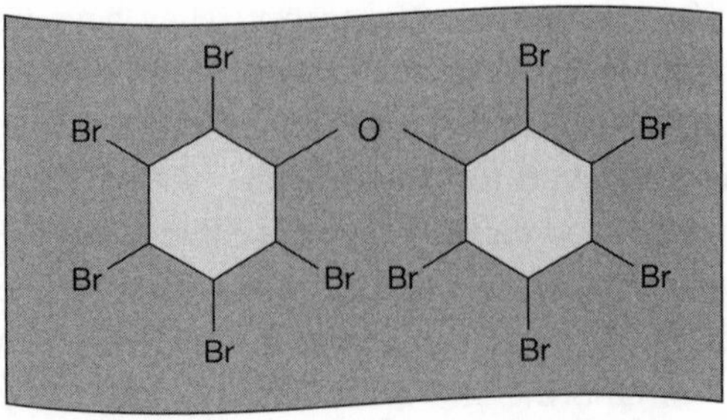

Figure 7.1 PBDE molecule showing all the positions where Br can be attached

most of the production has been in the United States. PBDE levels in tissues of people in the United States are 10 to 100 times higher than levels in Europeans and Asians. Canadians have somewhat lower levels than the United States, but surprisingly, children have higher levels than adults. These chemicals can cause long-term adverse effects in marine animals, and major reductions in reproductive success in fish and crustaceans.

What is the problem with fluorinated compounds?

Fluorinated compounds are also of concern. Perfluorinated compounds (PFCs) are a family of man-made chemicals that are used to make products that resist heat, oil, stains, grease, and water. Common uses include nonstick cookware, stain-resistant carpets and fabrics, coatings on some food packaging (e.g., microwave popcorn bags and fast food wrappers), and fire-fighting foam. These chemicals, such as perfluorooctanesulfonic acid (PFOS) and perfluorooctanoic acid (PFOA), are persistent and ubiquitous in the environment. They are also likely to be toxic and bioaccumulative. They are, like DDT, PCBs and dioxin, halogenated. Halogens include fluorine (F), chlorine (Cl), and bromine (Br). Halogenated compounds, with either F, Cl, or Br in their structure are resistant to microbial degradation. Perfluorooctanesulfonic acid has been detected by Jessica Reiner and colleagues in tissues of marine mammals from Arctic waters, suggesting widespread global distribution. PFOS in nursing Hudson Bay beluga whale calves exceeds the oral reference dose (the level considered safe for humans), which raises concern for harmful effects in sensitive Arctic marine wildlife. One wonders why, when there was so much concern and attention to chlorinated chemicals (DDT, PCBs) decades ago, people didn't realize that brominated and fluorinated chemicals would be likely to have similar behavior in the environment and similar effects and look into them.

What is the concern about alkylphenols?

Alkylphenols are chemicals used in the production of detergents and other cleaning products, personal hair care products, and commonly used plastics. They are also known to be endocrine disruptors, specifically estrogen mimics. One alkylphenol, bisphenol-A (BPA), has been the subject of concern and controversy regarding its potential adverse effects on human health, particularly children's development, and until recently it was commonly used in plastic baby bottles and other products. In 2008 Canada banned the use of BPA in baby bottles. Most scientists who study alkylphenols consider them serious environmental hazards with hormone disruptive effects in humans and wildlife, including marine animals. For example, scientists have found extensive contamination in lobsters in urban areas, and that alkylphenols are toxic to them at low doses, interfering with metamorphosis and shell hardening. Reproductive and developmental effects of BPA in fishes include decrease of male hormones, death of testicular cells, inhibition of sperm and egg production, and decreased hatchability of larvae.

What are nanoparticles and what is the concern about them?

Nanotechnology is used in many areas of modern life, including the manufacture of paints, batteries, fuel additives, catalysts, transistors, lasers, lubricants, medical implants, water purifiers, sunscreens and cosmetics, and food additives. Nanoparticles are microscopic particles that are larger than individual molecules, and have at least one dimension less than one hundred nanometers (a nanometer is one billionth of a meter or one millionth of a millimeter). Nanomaterials or nanoparticles (NPs) (<100 nm) can be made of different materials; some come from combustion like diesel soot, and some are manufactured. Because of their size, they have unusual properties that make them useful for drug delivery, gene therapy, and other biomedical uses, as well as in the optical, cosmetics, materials science, and electronics fields. They may be made of carbon (nanotubes, fullerenes), transition metals (gold, platinum, silver), metal oxides (titanium dioxide, zinc oxide), plastic (polystyrene), or silica, and are being manufactured in increasing amounts. Fullerenes, named after Buckminster Fuller (the designer of geodesic domes), are hollow spherical molecules composed of 60 atoms of carbon. Informally called buckyballs, they resemble soccer balls.

One reason for concern about nanomaterials is that since they are so small, they may interact with the environment and living things in unexpected ways. They are extremely diverse, exhibiting a wide variety of properties. Particular classes are of concern because of their potential impacts on human and environmental health, including nanosilver, carbon nanotubes, and fullerenes. NPs pose possible dangers because they are reactive and can pass easily through cell membranes. They can cause inflammation in the lungs, and because of their tiny size they are highly mobile and able to move from their original site (the lungs from being inhaled) to other parts of the body. Inside cells, NPs can stimulate the formation of reactive oxygen species (ROS) that interfere with DNA, proteins, and cell membranes.

Greater use of NPs has led to their release into the environment in runoff and sewage effluent, and their accumulation in coastal environments. They have come under scrutiny as potential pollutants. For example, the nanoparticle form of titanium dioxide after exposure to ultraviolet radiation can be toxic to marine life. A field of nanoecotoxicology is developing. Investigating effects of NPs in the aquatic environment is important, since it receives runoff and wastewater from domestic and industrial sources containing nanoparticles. While metal NPs may have fates similar to other forms of the same metal, metal NPs tend to be more toxic than regular forms of the metal. However, metals in NPs may be tightly bound to the core material and not readily dissociated. In the aquatic environment NPs tend to agglomerate, which means they will settle into sediments and be taken up by organisms. Danielle Cleveland and colleagues compared the environmental fate of nanosilver in consumer products, two silver (Ag) NP standards, and ionic silver (Ag$^+$) in estuarine mesocosms containing a variety of species, and found that the consumer product (a stuffed teddy bear) released high amounts of Ag (>80%) over 60 days, which moved from the water into clams, grass shrimp, mud snails, cordgrass, biofilms, and sediment. Ag was initially adsorbed from the water onto the sediment, then from there moved into the clams and other residents in the tank. Significant amounts were taken up into animals by consuming sediments and smaller organisms.

Research is ongoing to develop methods to measure NPs in water and sediment, and to determine their environmental occurrence, the sources and pathways of their release, their transport and fate, and their potential effects. There is a need to develop standardized analytical techniques, understand the role of wastewater treatment plants on their environmental fate, and determine mechanisms of their transport and fate in the environment. This will be difficult, since so many chemicals can be in the nanoparticle form, but will not be able to be measured in the same way.

NPs accumulate in estuarine organisms, with effects largely unknown. A limited number of studies have shown toxic effects, but the effects are highly specific to the chemical nature of the NP and the organism. Fullerenes and nanotubes produced adverse effects on fish, and metal NPs caused deleterious effects in several fish species and invertebrates. Because of limited data, scientists and regulators have been reluctant to propose broad guidelines limiting their use. This is an example of something becoming widespread in the environment before we have learned much about its effects.

Are existing regulations adequate to protect against harm to marine life, wildlife, and humans by these new chemicals? Are there any technological improvements?

The current regulatory framework cannot keep up with the fast pace of new chemical development and new uses. It seems that chemicals get regulated in the US only after they have become widespread, have been proven to be harmful, and after they have caused extensive damage. That is a basic failing and weakness of the toxic substance law, the Toxic Substances Control Act (TSCA), which currently seems to protect the chemical manufacturers to a greater degree than the environment or human health. New methods of measurement and better review of the ecological risk of new chemicals is needed. Undoubtedly additional types of pollutants that we know nothing about will continue to be found once we know to look for them.

However, research in 2013 found out what happens to nanosilver in a wastewater treatment plant—it does not remain in metallic form for long, but is transformed into a silver sulfide salt. This is good news, because the silver sulfide salt causes much fewer problems because this form of Ag is much less soluble. In sewage treatment plants, about 95% of the nanoparticles were bound in the sewage sludge, leaving only 5% in the treated wastewater. This percentage could be further reduced by using better filters.

What is Noise pollution?

Noise pollution in the ocean is another emerging concern. For millions of years, the oceans have been filled with sounds from natural sources such as the clicks and songs of whales, and the grunts, croaks, and drumming of fishes named after their sounds. Many marine species have acute hearing, echolocation, and communication abilities. Time was when a blue whale call could be heard by others of its species many miles away, but since the advent of the propeller engine 150 years ago, that has changed. An increase in boats, commercial shipping traffic, exploration and extraction of oil and minerals, air guns used for seismic exploration, sonar, and even jet skis contribute to the increased level of underwater noise. Sound travels four times faster in water than in air so it travels farther under water. High intensity sound can travel for thousands of miles. Since water is denser than air, sound waves travel though water at higher energy levels and are therefore louder.

What types of noise occur in the ocean?

Underwater noise has been divided into two main types: (1) impulsive—loud, intermittent or infrequent noises, such as those generated by pile driving and seismic surveys; and (2) continuous—lower-level constant noises, such as those generated by ship engines and wind turbines. These two types of noise have different impacts on marine life. The frequency or pitch of the noise is also important, as animals are sensitive to different frequencies. For instance, most of the noise produced by pleasure boats is low frequency, below 1.5 kilohertz (kHz). Although most sensitive to sounds above 15 kHz, bottlenose dolphins could be disturbed by these boat noises because they hear in the wider range 0.075–150 kHz and some of their calls are below 2 kHz.

One new noise source having immediate and obvious negative effects has been the development and testing of low-frequency active (LFA) sonar that has a potential worldwide deployment by the US Navy. Several tests of this sonar have resulted in deaths of many marine animals. The oil and gas industry uses arrays of airguns which release intense impulses of compressed air into the water about once every 10 to 12 seconds. Seismic surveys produce sounds with pressures higher than those of other man-made sources besides explosives. In seismic airgun testing, a ship tows a seismic airgun that shoots extremely loud blasts of compressed air through the ocean and miles under the seafloor to locate oil and gas deposits. These airguns must be incredibly powerful in order to penetrate the water and the earth's crust and then bounce back up to the surface. In fact, this sound is 100,000 times more intense than a jet engine. Tests would last 24 hours for 33 days and would kill or injure marine mammals (some of them endangered species), including whales, dolphins, porpoises, seals, and otters. This powerful disturbance drives whales, dolphins, sea turtles, and fish away from feeding or fishing grounds, disrupts important behaviors like mating or caring for their young, and could cause permanent hearing loss or death. A deaf marine mammal cannot survive as they rely on hearing to communicate, navigate, and find food. There is no way of knowing if an impact kills an animal outright or whether it dies later from hearing loss or environment impact. Seabirds and other species, such as endangered sea turtles, could be affected as well. Fishermen are concerned about detrimental effects on their livelihoods.

What effects are produced by noise pollution?

Noise pollution can cause both lethal and sublethal effects, as reviewed by Weingart. Most animals are alarmed by the loud sounds, which may damage internal organs (especially ears), and cause panic. Normal communication between marine animals can be disrupted by noise. Scientists are investigating which frequenc lies and at what levels noise negatively affects marine life. Sea mammals have excellent hearing to take advantage of sound and to compensate for poor visibility—the heads of whales and dolphins are full of resonant chambers that give them extraordinary hearing. Naval sonar is comparable to bomb explosions. The navy estimated that approximately 2,000 cetaceans (whales and dolphins) died from exposure to sonar and more than 5 million suffered some degree of hearing loss after training exercises. The navy estimates that even from 300 miles away, a sonic blast is 100 times stronger than marine mammals can withstand. Deaths of animals, especially cetaceans, often occur hours after exposure to extremely loud underwater noise. For example, whales die after beaching themselves shortly after a tactical sonar exercise; this is a rather common occurrence that has been reported in Greece, Madeira, Hawaii, Spain, and the coastal United States—areas where sonar exercises are common. In March 2000, at least 17 whales stranded themselves in the Bahamas, and a federal investigation identified testing of a US Navy active sonar system as the cause. Upon examination of the whale carcasses scientists discovered blood on their brains, ruptured ear canals, and bubbles in their systems similar to when people contract decompression sickness (the bends). Beaked whales are the most common ones affected. Scientists attached digital devices to Cuvier's beaked whales off the coast of California to measure the noise they were exposed to and their response. When a simulated sonar signal was sounded at 200 dB and between 3 km and 10 km away, the whales stopped feeding and swimming, swam rapidly away from the noise, and some performed unusually deep and long dives.

Chronic effects of lower level exposures are also seen. Marine biologists have linked the loud noises to reduced vocalization, which suggests reduced communication, foraging, and breeding. Humpback whale song in the Stellwagen Bank National Marine Sanctuary was reduced during transmissions of an Ocean Acoustic Waveguide Remote Sensing experiment approximately 200 km away.

Other animals are also sensitive. Giant squid were found dead along the shores of Spain in 2001 and 2003 following the use of air guns by offshore vessels; autopsies indicated that the deaths were related to excessive sound exposure. A study of low frequency sound exposure by Andre and colleagues—similar to what the giant squid would have experienced—in four cephalopod species found that all of the exposed squid, octopus, and cuttlefish had massive acoustic trauma in the form of severe damage in their hearing organs.

Very loud, short, sounds, such as those produced during pile driving, can harm nearby fish. Organs most sensitive are those with gas/tissue interfaces (e.g., ears, swim bladders, air sacs). Fish with swim bladders are particularly susceptible to loud noises, such as from pile driving, because the gas in their swim bladders is expanded by sound pressure, which can cause the swim bladder to rupture. Other injuries include disruption of cells and tissues, internal bleeding, and auditory damage. However, more moderate underwater noises of longer duration, such as those produced by vessels, could potentially impact much larger areas and involve many more animals. For example, the foraging habits of chromis, a coral reef fish, were reduced due to boat noise, which also disrupted normal orientation and movements in larvae of cardinalfish. Playback of noise recorded from ships altered the feeding of sticklebacks and minnows, which consumed less food and showed startle responses.

There have not been many studies on effects of noise on other marine animals, but the few studies suggest widespread chronic effects. For example, crabs exposed to recordings of ship noise showed an increased metabolic rate, indicating elevated stress. Crabs also had reduced prey capture ability and were less able to escape predators when subject to loud noise. Effects were more severe on larger crabs than smaller ones.

What can be done about noise pollution?

While the use of sonar may be necessary in times of war, sonar training should not be held in areas inhabited by cetaceans. Training personnel about whale migration patterns would reduce unnecessary harm to these animals. In March 2013, the California Coastal Commission rejected a US Navy explosives and sonar training program planned off the Southern California coast that critics said could harm endangered marine life. The commissioners ruled that the navy didn't have enough information to support its claim that the threat to marine mammals would be negligible, and were concerned that the increased sonar activity could harm endangered animals such as the blue, fin, and beaked whales. While the navy estimated that 130 marine mammals could die and another 1,600 could lose hearing from the program (which plans over 50,000 explosions and 10,000 hours of high-intensity sonar use annually), critics considered this a gross underestimation because the area encompasses 120,000 nautical square miles off the coast of Southern California, including a corridor between it and Hawaii, waters used by many endangered cetacean species. The commission wants the navy to create safety zones with no high-intensity sonar activity near marine sanctuaries, protected areas, and areas with high seasonal concentrations of blue, fin, and gray whales.

The National Oceanic and Atmospheric Administration (NOAA) is documenting human-made noises in the ocean and turning the results into large sound maps, which use bright colors to symbolize the sounds in the oceans. The scale goes from red (115 decibels at the top) to orange and yellow, and then to green and blue (40 decibels at the bottom), and presents the results in terms of annual averages rather than peaks. Many areas of the ocean surface (where whales and other marine mammals spend most of their time) are orange, indicating high average levels. The project's goal is to better understand the nature of the noise and its impact on mammals. The maps are enabling scientists, regulators, and the public to visualize the serious risk that noise poses to marine life. The findings are likely to prompt efforts to reduce the problem through laws, regulations, treaties, and voluntary noise reductions, nationally and internationally. However, one might question the validity of using annual averages rather than the maximum—if there is very loud noise for one week and none during the rest of the year, the average will be very low, but many animals could be killed or maimed during that one week.

Many vessels used for fishing and research are being designed to create less noise. The government already has some authority to regulate oceanic sound in United States waters through the Endangered Species Act and the Marine Mammal Protection Act. The International Maritime Organization (IMO) of the United Nations also has the authority to set acoustic standards. In the past few years, it began discussing how to achieve voluntary noise reductions. Since many commercial vessels are foreign and most shipping noises are in international waters, IMO's backing is important for reductions to be substantial enough to be effective. Ships can be built that are far quieter than those in use. These new designs will also be more efficient and pollute less. It would be very expensive, however, to retrofit existing ships.

In June 2013 a coalition of conservation groups (the Natural Resources Defense Council, the Gulf Restoration Network, the Center for Biological Diversity, and the Sierra Club) announced a major agreement with the Department of the Interior and oil and gas industry representatives to protect whales and dolphins in the Gulf of Mexico from high-intensity airgun surveys by the oil and gas industry. The settlement requires new safeguards, including putting biologically important areas off-limits, expanding protections to additional at-risk species, and requiring the use of listening devices to better ensure that surveys do not injure endangered sperm whales. In November 2013, NOAA Fisheries announced final regulations requiring the US Navy to implement protective measures during training and testing activities in the Atlantic Ocean and Gulf of Mexico to reduce effects on marine mammals. The measures include establishing marine mammal mitigation zones around each vessel using sonar; using navy observers to shut down sonar operations if marine mammals are seen within designated zones; ensuring that explosives are not detonated when animals are detected within a certain distance; implementing a stranding response plan with a shutdown provision in certain circumstances; allowing the navy to contribute in-kind services to NOAA Fisheries if the agency has to conduct a stranding response and investigation; and using specific mitigation measures to reduce effects on North Atlantic right whales.

Are there concerns about radioactivity in the marine environment?

Concerns about radioactivity were greatly reduced after the nuclear test ban treaties several decades ago. Nevertheless, there have been many incidents in which radioactive material has been dumped or discharged into the oceans, accidentally or on purpose. A British nuclear fuels plant has repeatedly released radioactive waste into the Irish Sea, a French nuclear reprocessing plant has discharged radioactivity into the English Channel, and the Soviets dumped large amounts of radioactive material into the Arctic Ocean and Barents Sea. However, it took the meltdown of the Fukushima plant in Japan following the March 2011 earthquake to reawaken concerns about radioactivity in the ocean, including levels in fish near Japan as well as in migratory fish that could carry radioactivity with them across the ocean. Both short-lived radioactive elements, such as iodine-131 (with a half-life of eight days) and longer-lived elements such as cesium (Cs)-137 (with a half-life of 30 years) can be absorbed by phytoplankton, zooplankton, kelp, and other marine life and move up the food chain into fish, marine mammals, and humans. Studies show that radioactive material moves with ocean currents, is deposited in marine sediments, and moves up the marine food web. Once taken up by the body it gets into the bloodstream, from which it is deposited in bones and other tissues, potentially causing genetic damage or cancer. Depending on the chemical form which organisms take up, radiation may also concentrate as it moves through the food chain.

The wrecked Fukushima power plant released very high amounts of radiation into the Pacific, where cesium levels shot up to 45 million times the background level. Thousands of tons of radioactive water were released into the ocean, and smaller amounts continue to be released. In June 2012, 15 months after the accident, 56% of fish tested by the Japanese government were contaminated with cesium-137 and Cs-134, products of nuclear fission. Over 9% of the fish catches exceeded Japan's official ceiling for Cs. Radiation levels were high in many species that Japan exports, such as cod, sole, halibut, carp, trout, and eel. Tuna, octopus, and anchovies have declining Cs levels after much higher contamination in the months just after the accident. However, 69% of anchovies still had some Cs contamination in June 2012, as did 32% of the tuna. Commercial fishing has been banned along the Fukushima coastline, although the discovery of contaminated fish outside the region prompts concerns that the radiation has spread farther away. Tuna can migrate across the ocean, and of 15 Pacific bluefin tuna caught off the California coast in 2012 and analyzed by Madigan and colleagues, all had radioactive Cs. Although levels were well below the standard, this confirms that radiation from the disaster has been carried around the world by migrating fish. In March 2013, additional information about radioactivity in fish near Japan revealed a continued cause for concern. A newspaper reported that one greenling registered very high levels, as did a rockfish. These were individual fish and not a representative sample, but nevertheless, the levels were very high. Reports of continued leakage of highly radioactive water continued in 2013.

What is light pollution?

Light pollution is excessive or obtrusive artificial light. Like any other form of pollution, it can disrupt ecosystems and have adverse health effects. The most well-known marine effect of nighttime lighting is the disorientation of hatching sea turtles, which emerge from nests on beaches at night and need to orient to the ocean. They wait just beneath the sand surface until conditions become cool. This temperature cue prompts them to emerge primarily at night, although some emerge in the late-afternoon or early-morning. They find the ocean by moving away from the dark silhouette of dunes and their vegetation. Sea turtle hatchlings have an inborn tendency to move in the brightest direction. On a natural beach, the brightest direction is the open view of the night sky over the ocean. Hatchlings also tend to move away from darkly silhouetted objects associated with the dune. This behavior can take place during any phase and position of the moon, which indicates that they do not depend on lunar light to lead them to the sea. They become disoriented by the brightness and glare of artificial lights from hotels, condominiums, and other buildings near the beach. To a hatchling, an artificial light appears bright because it is relatively close by, but not intense enough to brighten the sky. The glare makes the direction of the source appear much brighter than the other directions, so they will move toward the artificial light no matter where it is relative to the sea. While crawling the wrong way on the beach, hatchlings exhaust their limited energy stores, which they need once they reach the ocean where they must swim out as far as 60 miles offshore toward the floating *Sargassum* seaweed. Disoriented hatchlings may wander inland, where they can die of dehydration or predation, or may be run over or drown in swimming pools. Artificial lighting causes thousands of hatchling deaths each year in Florida alone, and is a significant marine turtle conservation problem.

What can be done about light pollution?

Reducing the amount of artificial light that is visible from nesting beaches is the first step to reducing light pollution. Coastal communities around the world have passed laws that require residents to turn off beachfront lights during turtle nesting season. At Gulf Islands National Seashore, about half the nests had a high level of hatchling disorientation. But after an education program, there was a 6% reduction in just one year, showing how educating the public about light pollution can benefit sea turtles. There are also new types of lighting fixtures, red or amber lights, which are considered sea turtle-friendly because turtles do not detect these wavelengths readily.

8

BIOACCUMULATION AND BIOMAGNIFICATION

What is bioaccumulation and what is biomagnification?

Organisms take pollutants up from the environment through their skin, gills, or digestive system. The term bioaccumulation is generally used to describe uptake, but there are specific terms that refer to specific ways they do so. Bioaccumulation refers to uptake from all sources in the environment. The bioaccumulation factor (BAF) refers to the concentration of the chemical in the organism compared to that of the sediment, when that is the major source of uptake. Bioconcentration is a more specific term that refers to uptake from water. The bioconcentration factor (BCF) is the concentration of a chemical in the organism relative to that in the water. Biomagnification refers to increasing levels of a contaminant from one trophic level to the next in a food chain (Figure 1.3), due to accumulation from food (trophic transfer). The biomagnification factor (BMF) is the concentration in a species at one trophic level divided by that at the trophic level below (the food of the species in question). Warmer temperatures cause greater bioaccumulation, probably because animals need to eat more due to their elevated metabolism.

What happens once a metal is taken up into an organism?

Once a chemical is taken up, the organism may store it, metabolize it, or eliminate some of it through gills or urine. Metals can't be metabolized, so they tend to be stored to varying degrees. Essential metals, for example copper (Cu) and zinc (Zn), tend to be regulated to some optimum concentration, above which the animal will excrete the excess. Potentially toxic metals must either be excreted or stored in a nontoxic form if they are not to cause damage. Toxicity occurs when the concentration exceeds the amount that can be stored in these nontoxic forms or excreted. Metals tend to be stored in specific tissues such as the liver, which generally has the highest concentration of Cd, Zn, Cu, Pb, and Cr. However, significant levels may also be found in muscle, which has implications for human consumption of edible species. For example, some edible crustaceans, such as blue crabs, store Hg largely in muscle. Since crustaceans must periodically molt their shell, depositing metals in the shell prior to molting it is a useful way to get rid of contaminants. In corals, the symbiotic algae that live in the coral tissue (zooxanthellae) accumulate greater metal concentrations than the skeleton or living coral tissue itself. The loss of zooxanthellae during stress (bleaching) may help reduce metal levels in the corals. Mollusks secrete a lot of mucus when stressed, which is also a way to get rid of metals. Fishes take up metals from water or food, but the type of food influences the degree of trophic transfer to the fish. Most of the Hg in fish tissues is in the form of meHg, which biomagnifies through the food chain and accumulates over time, reaching highest levels in old large carnivorous fish.

Where and how are metals stored in organisms?

The location inside a cell where metals are placed strongly affects their toxicity since metals associated with sensitive organelles and enzymes can impair cell functioning. Metals tend to bind to proteins and may prevent the protein from functioning normally. For example, metals can bind to active sites of essential enzymes, which are proteins, changing their shapes and inhibiting their activities. However, there are special proteins (metal-binding proteins, e.g., metallothioneins, MT) that can bind metals and make them unavailable. MTs are low molecular weight, heat-stable proteins that bind high amounts of metals and prevent the metals from doing damage. There is a limit to how much the MTs can store, however. Some animals, such as crustaceans, can also store metals in the form of insoluble metal-rich granules or deposits in tissues. Thus, metals may be toxic and available or may be unavailable, depending on where they are located and what they are bound to. In those animals that can accumulate high concentrations of metals without major effects, most of the metal is in nonavailable form such as MT and granules, which may explain their ability to survive in highly contaminated environments. The site of storage can also affect how much metal will get trophically transferred to predators, as found by William Wallace and Samuel Luoma. A predator would be better off eating prey with its metals tied up in granules rather than bound to MT, which is more trophically available to the predator.

How are organic contaminants taken up by organisms?

After taking up foreign hydrocarbons organisms may metabolize, store, and/or excrete them. Small zooplankton take up organic contaminants from the water, while benthic species accumulate them mostly from sediments, which generally have higher levels than the water. Concentrations of chlorinated organics (e.g., PCBs, DDT) in mollusks can exceed those in nearby sediments tenfold. Chlorinated organic chemicals tend to be metabolized slowly if at all, and are bioaccumulated and stored in the liver (or hepatopancreas, the comparable organ in invertebrates), or in blubber of marine mammals. These chemicals accumulate in fatty tissues, including yolk and liver, and biomagnify up the food web. In general, the most important factor determining an animal's concentrations of these chemicals is the trophic level, followed by the lipid content of the animal. Therefore, fatty predatory fishes (e.g., bluefish) will have higher concentrations than less fatty fish (e.g., striped bass) at the same trophic level in the same location. Chemicals tend to increase with the age and size of the animal. For animals high on the food web, food is the major source of uptake. Fish can accumulate PCBs directly from sediments and food. The amount of accumulation may be affected by the prey, the magnitude of contamination, movement patterns, trophic level, growth rate, and length of exposure (i.e., age). Female fish are able to eliminate some of their PCBs by putting them into the eggs (yolk is very rich in lipids to which PCBs bind). This accounts for observations such as those of Haim von Westernhagen and colleagues of female fish having lower PCB levels than males. While it is good for the female to reduce her PCBs, it doesn't seem like a good long-term strategy for the sensitive early life stages of the next generation to start out life with a built-in dose of these chemicals. Marine birds and mammals, high in food webs, accumulate high concentrations of PCBs and chlorinated pesticides. Walruses and whales in the far north, far from any use of such chemicals, also have high concentrations. Similar to the situation of fish with eggs, nursing mammals pass high levels on to their babies, which then start off their lives with elevated body burdens of toxic chemicals.

How do organisms metabolize organic contaminants?

PAHs (polycyclic aromatic hydrocarbons), which are accumulated primarily in the liver, can be transformed into chemicals that can be excreted via the gills and kidneys. There is an enzyme system that breaks down these chemicals. The enzymes responsible for oxidation of foreign compounds are called mixed function oxidases (MFOs), which include the highly studied cytochrome P-450 (CYP) system. Found in many organisms and tissues, CYPs are involved in metabolism of a wide range of organic compounds including PAHs,

PCBs, pesticides, and other chemicals. The reactions have two steps. Phase I reactions hydrolyze or oxidize the molecule in order to make it more water soluble. Phase II reactions involve combining the product of phase I with a substance that makes it less bioactive and more readily excreted. Sometimes, products of phase I are more toxic than the original pollutant.

Highly chlorinated compounds are metabolized very slowly, so they tend to accumulate. Lesser chlorinated compounds can be more rapidly metabolized and eliminated. Organophosphorus pesticides can also be oxidized by the MFO system.

Which marine organisms are good sentinels for bioaccumulation and biomagnification of chemicals?

Sedentary bivalve mollusks like mussels are often used for monitoring pollutants in local areas. They tend to be relatively low on the food web and not accumulate very high levels. In the oceans, mammals such as seals and cetaceans are useful sentinels for pollution. Their blubber accumulates high levels and can be sampled without harming the animal. Seabirds are also useful, and can be cheaper and easier to sample. They range widely across oceans, feeding as they move, but return every year to breed in the same location. In a seabird colony biologists can sample blood, feathers, oils, and biopsies, which can provide information on pollution at different spatial and time scales. Bird eggs, rich in lipids (fats), are excellent accumulators of fat-soluble chlorinated organic chemicals that are of most concern. Birds with different feeding habits can be sampled—for example, cormorants that forage mainly on fish in near-shore environments, auks that feed on smaller fish and zooplankton on the continental shelf, and pelagic seabirds such as storm petrels that range offshore, feeding on zooplankton and larval fishes. They all return to breed in colonies, where they can be sampled for contaminants and tagged for tracking. Seabird egg monitoring documented the decline of the persistent organic pollutants such as DDT in coastal habitats by the early 1980s, but DDT persisted at lower concentrations at that time. However, even DDT may now be finally disappearing.

Many pollutants are adsorbed onto plastics in the ocean, which are found in seabird stomachs (see Chapter 3). Species that eat a lot of plastic also have elevated contaminant levels and may experience toxic effects. In the past decade, other contaminants have emerged that are similar to chlorinated organic chemicals but have bromine or fluorine instead of chlorine (see Chapter 7). Eggs archived in specimen banks have enabled scientists to analyze the history of such pollutants of emerging concern.

What are safety issues for humans who consume seafood that may be contaminated?

Chemical contaminants in fish and other seafood may pose a potential health hazard to people who eat them over a long time. The hazard is from long-term exposure rather than a single exposure, (for example one meal) which might be the case with food that had microbial contamination. Fish are harvested from waters with varying levels of industrial chemicals, pesticides, and metals, which may accumulate to levels that can harm human consumers. There is concern about these contaminants in fish from freshwater, estuaries, and near-shore coastal waters more than from the open ocean. Pesticides used near or in aquaculture operations may also contaminate both wild and farmed fish. Federal safe levels and guidance levels are established for some of the most toxic and persistent contaminants found in seafood. States often use these levels to decide whether to issue advisories or to close waters for commercial harvesting of all or certain species of fish. Fish can accumulate inorganic chemicals including arsenic, cadmium, lead, mercury, selenium, copper, zinc, and iron—of which the most concern is mercury, which is the most toxic. In the case of mollusks, agencies use the degree of chemical contamination as part of the classification of harvesting waters; they allow harvesting from some waters, not from others, or only at certain times or under certain conditions.

What is Minamata disease?

The first example that made the world aware that the accumulation of contaminants in edible seafood could harm humans was in Minamata, Japan. This community had a factory that used Hg in the production of plastics and discharged Hg into the nearby bay, from which the people ate fish that had accumulated it in their tissues. From 1932 to 1968, the Chisso Corporation dumped an estimated 27 tons of Hg compounds into Minamata Bay. The Hg became methylated in the sediments of the bay, and then biomagnified up the food web. Methylmercury, the form that is especially toxic and biomagnifies in food webs, is produced by bacteria in the environment (Figure 5.1). Thousands of residents developed severe neurological and developmental defects, a condition now called Minamata disease, alerting the world that exposure to Hg can cause permanent behavioral and neurological effects. Severe cases led to insanity, deformation, and death. Many children whose mothers had eaten contaminated fish while pregnant were born with major disabilities. Congenital Minamata disease was observed in babies born to affected mothers, but also to mothers who did not have severe problems. These babies had symptoms of cerebral palsy. Affected people had numbness in their limbs and lips, slurred speech, and impaired vision. Some people had serious brain damage. Even before the symptoms appeared in the people, the cats in the village showed symptoms of Hg poisoning. The cats, which ate scraps of fish from fish markets and the table, died with symptoms similar to those only later seen in humans. People initially thought the cats were going insane when they witnessed their odd behavior. This led researchers to believe that the outbreak was caused by some kind of food poisoning, with contami-nated fish and shellfish being the prime suspects. Hg in the fish was finally linked to the disease. Hair samples were taken from affected people and from the Minamata population in general. In patients the maximum Hg level was 705 ppm (parts per million), indicating very heavy exposure. In unaffected Minamata residents the level was 191 ppm, compared to an average level of 4 ppm for people living outside Minamata.

Are there any concerns about mercury pollution in seafood today?

While gross pollution such as in Minamata is a thing of the past, there are still two major concerns. One is eating fish from mercury-contaminated areas. Agencies in each state moni-tor fish for the presence of contaminants and alert the public through bans (closures) and advisories when a threat to human health may occur from the consumption of contaminated fish. In waters with bans, possession and consumption of fish and/or shellfish is prohibited. An advisory is a recommendation to limit consumption to specified quantities, species, and sizes of fish. Those areas that are known to have elevated Hg will have commercial fishing prohibited and have warning signs for recreational fishers to not eat the fish. It is not known how many anglers ignore these signs in favor of a free meal (with no extra charge for the mercury).

The second major concern is the buildup of Hg in large car-nivorous fishes that are not from contaminated areas, but are high in the food web. This Hg comes mostly from atmospheric sources, especially from burning of coal in power plants. While other industries have had considerable reductions in emissions, Hg pollution from electric utilities is still of con-cern. The Hg released from coal burning power plants near coastal areas goes into the air and eventually comes down into the ocean. Over the past century, Hg in the surface ocean has more than doubled. Nearly all fish and shellfish contain traces of meHg. However, large carnivorous fish that live longer have the highest levels, because they have had more time to accumulate it. These large fish (swordfish, shark, king mackerel, and tuna) pose the greatest risk.

The greatest concern about Hg exposure is for a developing fetus. As seen in Minamata, some women who showed no signs of poisoning gave birth to children with severe brain damage. This is because meHg readily crosses the placenta. It can also be passed through breast milk to infants. This is of particular concern, because young children are more susceptible to Hg toxicity and the brain may be more affected as it develops. There is a correlation between prenatal exposure to Hg and decreased performance of infants and children on neurobehavioral tests including tests of attention, fine motor function, language skills, visual-spatial abilities, and memory.

While the danger of Hg poisoning may seem like a good reason to avoid consuming fish, the benefits of eating fish may outweigh many of the risks. Fish are high in protein, low in saturated fats, and contain important nutrients such as heart healthy omega-3 fatty acids. Eating fish reduces the risk of heart attacks, lowers blood pressure, and improves arterial health. So it is a matter of choosing the right fish to eat.

The US Environmental Protection Agency (EPA) and the US Food and Drug Administration (FDA) have issued consumption advisories for certain groups of people. They advise women who may become pregnant, pregnant women, nursing mothers, and young children up to age six is avoid fish high in Hg and limit the amount of fish consumed each week. They advise them not to eat shark, swordfish, king mackerel, or tilefish at all because they contain high levels of Hg (>1 ppm), but to eat up to 12 ounces (two average meals) of fish and shellfish that are low in Hg per week. Children should only eat six ounces of fish. Low Hg fish and shellfish include shrimp, canned light tuna, pollock, salmon, and tilapia. Albacore tuna contains moderate amounts of mercury. The EPA and FDA advise eating only 6 ounces of albacore tuna a week and advise the public to check for local advisories on fish caught from local waters that may be more greatly affected by pollution sources. These guidelines are not meant for adult men or for woman past childbearing age, but individuals concerned with exposure to Hg should follow them as well. The best solution may be to refrain from eating the large fish species often and to focus on eating smaller fish that do not have high levels of Hg. Cod, salmon, haddock, herring, and sardines, for example, do not have high levels.

Can metal pollution be found in calcium supplements derived from oyster shells?

Calcium is needed for prevention or treatment of osteoporosis, but some calcium supplements have been found to have elevated metals. Lead (Pb) has been found in some calcium carbonate supplements labeled as oyster shell. Pb is a neurotoxin that affects the brain and nervous system. The level of contamination has decreased recently, but still may present a health risk. Calcium supplements rarely list their Pb content, which should be less than 2 parts per million. California scientists analyzed the Pb in a variety of calcium supplements and found that two-thirds of them failed to meet the California's criteria for acceptable Pb levels in consumer products, which are stricter than the national criteria. Alternatives to so-called natural calcium supplements are plain calcium carbonate pills or calcium citrate.

What problems can result from eating seafood containing organic contaminants?

Persistant PCBs and related chemicals may remain a problem for quite some time. Because they become attached to particles in the water, they settle out in the sediments. When bottom dwelling animals feed, they ingest the contaminated sediments and pass them up the food chain, where they biomagnify and become most concentrated in carnivorous, fatty, large fishes. They tend to build up primarily in fatty tissue and to a less extent in muscle tissue. Scientists have found over a dozen different chlorinated organic compounds at higher concentrations in farmed salmon than wild salmon. PCBs are not highly toxic with a single dose (as in a single meal), but continued low levels of exposure (for example, eating contaminated fish over an extended period of time) may be harmful. There are standards set by EPA and FDA for safe levels of consumption. These numeric levels are based on lab tests of high concentrations on rats or mice followed by extrapolation downward to estimate what level would be safe in the rats or mice, and then extrapolation from rodents to people— so there is a large amount of uncertainly about these numbers, and no one should consider a concentration slightly above the standard to be alarming or a concentration just below it to be totally safe. The EPA considers PCBs to be probable human carcinogens, because they cause cancer in laboratory animals. Other tests on laboratory animals show damage from PCBs to the circulatory, nervous, immune, endocrine, and digestive systems. Risks to humans are highest in the fetus or nursing infant (as with Hg), when the mother is or has been exposed to PCBs. Women of childbearing age, especially those pregnant or nursing, are advised to minimize risk by avoiding eating fish from areas known to contain PCBs. In terms of chronic low-level exposure to PCBs over time, less is known about potential adverse health effects. However, scientists suspect that long-term exposure to small amounts can contribute to a variety of health problems including developmental problems in children, liver damage, and cancer. Some studies showed that children of mothers who ate fish from the Great Lakes with high PCBs had smaller head size, reduced visual recognition, and delayed muscle development. A mother's exposure to PCBs and other chemicals was linked to slight effects on her child's birth weight, short-term memory, and learning. Adults who ate fish containing PCBs and other contaminants had lower scores on several measures of memory and learning.

While none of these studies are a "smoking gun," they do provide reasons to avoid eating fish with high PCB levels. Since PCBs accumulate in fat, some procedures for preparing the fish can be useful. The amount of PCBs in fish may be significantly lower after cooking because lipids, along with lipophilic compounds like PCBs, tend to be removed from the fish during cooking. Before cooking one should remove the skin, the fat (found along the back, sides, and belly), internal organs, and the tomalley of lobster and the mustard of crabs, where these chemicals are likely to accumulate. When cooking, the fat should be drained away. Frying fish seals in the pollutants in the fish's fat, while grilling or broiling allows the fat to drain away. This can remove 20 to 30% of the PCBs. To smoke fish, it should first be filleted and the skin removed.

Organic chemicals associated with petroleum (PAHs) can also accumulate in seafood if it is exposed to the oil. The types and properties of oil influence whether seafood is contaminated. Crude oils and the products derived from them are complex and variable mixtures of hydrocarbons of different molecular weights and structures. Once exposed to oil, an organism becomes contaminated to the extent that it takes up and retains petroleum compounds. The BP oil spill of 2010 contaminated a very productive fishery with PAHs that accumulate in seafood, and are carcinogens and developmental toxicants. Seafood can be analyzed chemically for these contaminants, which is very time-consuming and expensive. Another way that seafood can be considered unfit for consumption, according to the National Oceanic and Atmospheric Administration (NOAA), is if it smells or tastes like petroleum; this is known as taint. A product tainted with petroleum is not permitted to be sold as food under US law. Petroleum taint in and of itself is not necessarily harmful, and may be present even when PAHs are below harmful levels. An open question is whether some fish could have higher levels of some PAHs that could not be detected by smelling it.

In response to the BP catastrophe, the FDA developed risk criteria and established thresholds for allowable levels of PAH contaminants in Gulf Coast seafood. Federal and state laboratories tested over 10,000 fish and shrimp for traces of certain PAHs from oil to be sure they were far below levels that could make anyone sick before commercial fishing was allowed to resume. However, some scientists, led by Miriam Rotkin-Ellman, disagreed with the levels that the FDA set because they failed to account for the increased sensitivity of fetuses and children. The scientists thought that the FDA also did not use appropriate seafood consumption rates, did not include all relevant health end points, and did not include protective estimates of exposure duration and acceptable risk. For two particular PAHs, benzo[a]pyrene and naphthalene, these scientists felt that safe levels should have been set far below the level that the FDA set, and that according to that lower standard up to 53% of shrimp samples were above levels of concern for pregnant women who eat a lot of seafood. It may be that the government was anxious to reopen the fishery sooner rather than later in order to reduce the already-devastating economic effects of the Deepwater Horizon catastrophe to the fishing industry in the Gulf.

Can dioxin contamination be found in seafood?

Dioxins and furans are among the most toxic chemicals, and they biomagnify up food chains. The amount of data on dioxins and dioxin-like PCBs in food is very limited and analytical measurements of these chemicals are difficult and very expensive. The greatest concentrations in food appear to be in freshwater fish. However some marine fish that are rich in lipids can accumulate worrisome levels. For example, 50 samples of Greenland halibut were analyzed for dioxins and dioxin-like

PCBs, of which 10 had concentrations that exceeded the EU's upper allowable limit. Atlantic halibut were analyzed for dioxins and dioxin-like PCBs, of which eight out of 14 belly samples showed very high levels of dioxins and dioxin-like PCB, exceeding the EU's upper allowable limit.

An industrial site in the Passaic River in Newark, New Jersey was contaminated with dioxin from the production of herbicides (Agent Orange) used in the Vietnam War. Dioxin is elevated in fish and blue crabs. There are warning signs posted in the area, yet many recreational anglers continue to eat the crabs they catch. A risk assessment done by the New Jersey Department of Environmental Protection suggests that people consuming whole crabs have a high risk of developing cancer. The site is on the Superfund list for pending cleanup, but the companies responsible for the cleanup suggested an alternative remedy—that instead of thoroughly cleaning up the whole river, they would start an aquaculture facility in Newark to grow clean fish, and station people along the river to trade their clean fish for the contaminated fish caught by fishermen. This plan, which ignored the fact that most of the catching and consumption was of crabs not fish, was criticized and ridiculed by environmental groups and in editorials in the local newspapers.

Can contaminants be found in fish oil supplements?

Since chlorinated organics concentrate in fish fat and oil, it is possible to find contamination in fish oil supplements. Large predatory fish like sharks, swordfish, tilefish, and tuna may be high in omega-3 fatty acids, but since they are at the top of the food chain, they also have high levels of persistent toxic substances. The Environmental Defense Fund (EDF) surveyed 75 companies that manufacture fish oil supplements and found that most supplements are adequately purified and safe. Consumers who take fish oil supplements should purchase them from companies that verified they have met strict standards for contaminants. California has stricter requirements than the FDA. Consumer Reports published a survey that revealed PCBs in amounts that could require warning labels under California's Proposition 65 (a consumer right-to-know law) in some of the supplements.

How can eating fish or shellfish that have accumulated HAB toxins cause disease?

Some single celled algae, both dinoflagellates and diatoms, produce toxins that can accumulate in the food chain and affect human consumers. Toxin-producing algae are normally found in the ocean at low concentrations and pose no problems. However, when they undergo a bloom, often in response to nutrients (see Chapter 2), it is called a harmful algal bloom (HAB). Filter-feeding shellfish pump water through their systems, filtering out and eating algae and other food particles. When they eat toxic algae, the toxin can accumulate in their tissues, often without affecting them much. Most cases of seafood poisoning are in people who ate shellfish that accumulated the toxins. When the bloom subsides, the shellfish eventually flush the toxin from their systems. In contrast with chemical pollutants that need to build up in one's system over a long time, when people eat a single meal or only a few HAB-contaminated shellfish, acute symptoms start shortly thereafter.

What is Paralytic shellfish poisoning (PSP)?

Paralytic shellfish poisoning (PSP) is a severe neurological condition caused by eating shellfish contaminated with saxitoxin, which is produced by the dinoflagellate *Alexandrium*. Blooms of *Alexandrium* are common in New England. Clams, mussels, oysters, and scallops can accumulate the toxin, as can some snails. Symptoms include tingling, numbness, burning, giddiness, drowsiness, fever, rash, and staggering. Effects generally last only a few days in nonlethal cases. The most severe cases can result in respiratory arrest within 24 hours of consumption, because the toxin paralyzes the diaphragm, making it impossible to breathe. PSP is prevented by large-scale monitoring programs (measuring toxin levels in shellfish) and rapid closures of toxic areas to harvesting of shellfish. In addition to measuring toxin levels in shellfish, predictions of blooms are based on the amount of *Alexandrium* in its cyst (dormant) stage detected in sediments the previous fall. In order to protect public health, shellfish beds are closed when toxicities rise above a certain level, often during the peak harvesting season. Due to effective monitoring by state agencies, there have been no illnesses from legally harvested shellfish recently, despite some major blooms. However, there have been some severe poisonings of individuals who ignored closure signs. The toxin is not destroyed by cooking the shellfish. Some shellfish can store the toxin for several weeks, but butter clams can store it for up to two years. PSP has been implicated as a cause of deaths of marine mammals such as sea otters after eating butter clams that accumulated saxitoxin. Ingestion of saxitoxin-containing mackerel was implicated in the deaths of some humpback whales. Deaths of fish including endangered sturgeon have also been associated with *Alexandrium* blooms.

What is diarrhetic shellfish poisoning (DSP)?

Diarrhetic shellfish poisoning (DSP), as its name suggests, causes diarrhea, although nausea, vomiting, and cramps are also common. Symptoms usually set in shortly after ingesting infected shellfish, and last for about one day. The toxin is okadaic acid, which causes intestinal cells to become very permeable to water, resulting in diarrhea with a risk of dehydration. The causative organism is the dinoflagellate *Dinophysis*, which is widely distributed. DSP is a significant problem in northern Spain, Ireland, and the Mediterranean/Adriatic Sea. The toxin has been detected in shellfish in Eastern Canada.

While no cases of DSP have been reported along the West Coast of the United States, *Dinophysis* is commonly found in British Columbia and Puget Sound in Washington State. As no life-threatening symptoms occur, no fatalities from DSP have been recorded.

What is neurotoxic shellfish poisoning (NSP)?

Neurotoxic shellfish poisoning (NSP) is caused by consumption of shellfish contaminated with brevetoxins primarily produced by the dinoflagellate *Karenia brevis.* Blooms of *K. brevis*, called Florida red tide, occur frequently along the Gulf of Mexico. Symptoms of NSP include gastrointestinal and neurological symptoms: nausea and vomiting; paresthesias (tingling sensation) of the mouth, lips, and tongue; and distal paresthesias, ataxia, slurred speech, and dizziness. Neurotoxic shellfish poisoning causes a mild gastroenteritis with neurologic symptoms comparable to paralytic shellfish poisoning. With the inhalation of aerosolized toxins, especially brevetoxins from sea spray exposure, respiratory irritation and other health effects occur in humans and other mammals. Neurological symptoms can progress to partial paralysis. Shellfish beds in Florida are routinely monitored for the presence of *K. brevis* and other brevetoxin-producing organisms. As a result, few NSP cases are reported from the United States. However, an alarmingly large number (several hundreds) of endangered Florida manatees were apparently killed by the toxins in 2013.

What is amnesic shellfish poisoning (ASP)?

Amnesic shellfish poisoning (ASP) is caused by domoic acid, which is produced by marine diatoms in the genus *Pseudo-nitzschia,* the first example of a toxin-producing diatom. When shellfish accumulate domoic acid in high concentrations during filter feeding, the toxin can be passed on to humans that eat them. Both shellfish and finfish can accumulate this toxin without apparent ill effects. The toxin can bioaccumulate in other phytoplankton eaters, such as anchovies and sardines. Domoic acid is a neurotoxin, causing short-term memory loss, brain damage, and death in severe cases. It has been responsible for several deaths and both permanent and transitory illness in over a hundred people. Amnesic shellfish poisoning was first discovered in late 1987, when a serious outbreak of food poisoning occurred in eastern Canada where a number of patients died and others suffered long-term neurological problems. Because the victims had memory loss, it was called amnesic shellfish poisoning. However, since the toxin has been found in finfish and the chemical structure of the toxin is now known, a more accurate term is domoic acid poisoning. It not only affects humans, but marine birds and mammals as well. Marine mammal and seabird strandings and deaths off the Southern California coast have been linked to this toxin. Most of the animals found dead, including sea lions and harbor seals, tested positive for domoic acid.

What is Ciguatera?

The most widely reported HAB toxin disease is ciguatera, which results not from eating shellfish but from consumption of contaminated reef finfish. It is estimated that at least 50,000 people per year who live in or visit tropical and subtropical areas suffer from ciguatera worldwide. Patients suffer for weeks to months with debilitating neurological symptoms. The dinoflagellate *Gambierdiscus toxicus* produces ciguatoxin (CGX) and similar toxins. The dinoflagellates are eaten by herbivorous fishes, which are then eaten by larger carnivorous fishes. The toxins move up the food web and concentrate in the fish. Larger individuals of species high up on the food chain in tropical and subtropical waters, such as barracudas, snappers, moray eels, groupers, triggerfishes, and amberjacks, are most likely to cause ciguatera poisoning, although other species may cause it. Ciguatoxin is odorless, tasteless, and heat-resistant, so fish cannot be detoxified by cooking them. Symptoms include gastrointestinal effects such as nausea, vomiting, and diarrhea, and neurological symptoms such as headaches, muscle aches, numbness, vertigo, and hallucinations. Severe cases can also produce a burning sensation on contact with cold. Symptoms have developed in otherwise healthy people after sexual intercourse with someone with ciguatera poisoning, showing that the toxin may be sexually transmitted. (However, it is not clear why someone with those symptoms would be in the mood for having sex.) Diarrhea and rashes have been reported in breast-fed infants of poisoned mothers, suggesting that the toxins get into breast milk. The symptoms last from weeks to years, and in extreme cases up to 20 years, often leading to long-term disability. Most people recover slowly over time, but some patients recover and then subsequently get recurring symptoms. Unlike beds of sedentary shellfish that can be monitored and closed when HAB toxins are found, fish are very mobile and the occurrence of ciguatera is very spotty. In addition, the *Gambierdiscus* don't need to bloom in order for fish to become contaminated. In a trawl full of fish caught at a given location, some may have ciguatera while others of the same species and size will not. Therefore it is very difficult to monitor it and reduce the occurrence of this debilitating disease. Although sensitive laboratory analyses can detect and confirm CTX in fish, no practical field tests are available for monitoring programs and detecting CTX in fish quickly enough before it would spoil. Prevention depends on educating the public, seafood suppliers, and distributors about known ciguatera areas and high-risk fish species. The only sure way to prevent it is to not eat fish when in the tropics—this is hardly a satisfactory solution to the problem.

How can the incidence of poisoning by marine toxins be reduced?

Ongoing surveillance and rapid detection are essential to reduce the incidence of poisonings. However, conventional sample collection at sea followed by analysis in a land-based laboratory is cumbersome and can take several days. One cannot wait that long before eating one's dinner. Some new technologies are available, including robotic environmental sampling processors (ESP) that use molecular probes to detect microorganisms in water and automated technology to provide near real-time information on what's in the water. The instrument was tested in Puget Sound in the summer of 2013 for its ability to provide early warnings of harmful algae, their toxins, and shellfish pathogens. Because the ESP can detect harmful algae and bacteria in the water in near real time, it can provide early warning of developing blooms before they contaminate shellfish. This information can help shellfish growers and public health officials make decisions to ensure safe seafood to protect public health.

9

CLIMATE CHANGE AND OCEAN ACIDIFICATION

What causes global warming or climate change?

The burning of fossil fuels emits carbon dioxide into the atmosphere, which results in the greenhouse effect—less heat can be re-radiated away from the earth, thus raising the temperature of the atmosphere and ocean. In the past century the oceans have warmed by about 1 degree F to a depth of 200 feet, and the overwhelming scientific consensus is that increasing levels of human-caused greenhouse gases in the atmosphere are the principal cause.

What problems are happening or expected to happen in the marine environment because of climate change?

Climate change is the biggest single threat to our oceans' health. The warming of the oceans will have numerous effects on all organisms, most basically elevating their metabolic rates, which ultimately affects life history, population growth, and ecosystem processes. Elevated metabolic rates create increased demand for oxygen at the same time that the warmer water can hold less oxygen. The uptake of toxic contaminants is also accelerated by elevated metabolic rates.

Variation in temperature can also affect the abundance and distribution of plankton. As the ocean's surface warms, it becomes more stratified—with greater temperature differences between warm surface water and cooler deeper water. Vertical water movements (upwelling), which bring nutrient-rich water up from deeper layers to surface waters (where most of the phytoplankton live), are reduced. Consequently, phytoplankton receive less nutrients and are less productive, because productivity requires nutrients. Upwelling can be stimulated by mixing due to winds. Less wind means less mixing, fewer nutrients for phytoplankton, and fewer phytoplankton to sustain fish populations. Concurrent with climate change, the annual primary production of the oceans has decreased since the 1980s. Modest changes in temperature have altered trade wind intensity in the Caribbean, reducing the supply of nutrients to phytoplankton and ultimately causing the collapse of some fisheries. Since late 1995, monthly observations of physical factors, including nutrient and chlorophyll levels and meteorological readings, have been collected at a site off the coast of Venezuela to establish a long-range record. The sea surface temperature increased about $1°$ C ($1.8°$ F) and winds decreased. But the effect on marine life was dramatic: populations of phytoplankton dropped, along with the local harvest of sardines. Changes in ocean currents caused by climate change could lead to shifts in regional climate and weather patterns.

Why are coral reefs particularly vulnerable?

Among the most sensitive groups of organisms are seagrasses, mangroves, salt marsh grasses, oysters, and corals, which all create habitat for thousands of other species. Current and future CO_2 levels will produce changes in ocean temperature and chemistry beyond what corals have experienced. Some scientists fear that conditions have already reached a "tipping point" for corals, which now are less able to recover from additional change. They are considered one of the most sensitive ecosystems to climate change, like the canary in the coal mine. Coral reefs have been in existence for over 500 million years, but their continued persistence is uncertain. With increases in ocean temperature, corals can bleach (Figure 9.1). Bleaching occurs when the corals lose their symbiotic single-celled algae, the zooxanthellae, which photosynthesize and provide food to the corals, and in turn receive protection and the nutrients needed for photosynthesis. Bleached corals appear white. Zooxanthellae are sensitive to stresses including temperature changes, and when they die or leave, bleached corals are usually unable to meet their energy requirements by filter feeding alone. In some cases, zooxanthellae return and the coral will survive. Coral death by bleaching and diseases due to increased heat and irradiation, as well as decreased calcification caused by ocean acidification (discussed later), are among the most important threats. Since the 1980s, major bleaching events have increased around the globe—for example, in 1998, 80% of the coral reefs in the Indian Ocean bleached, causing 20% of them to die.

Reef recovery is thought to depend on arrival of larvae from distant, interconnected reefs. Observations of relatively rapid recovery of corals following a mass bleaching event suggests

Figure 9.1 Coral Bleaching (photo from NOAA)

that corals can recruit from local sources, especially in the absence of human-caused disturbances, which slow down recovery.

Scientists have found an early warning sign for corals that may bleach—some proteins in the zooxanthellae respond rapidly and dramatically to temperature stress. Before actual bleaching, hemoglobin genes are expressed at a higher level. Because of this sensitivity, hemoglobin production by the algae may be able to be used as an early warning indicator of stress. Scientists have also found some heat-resistance genes that enable corals in some areas to avoid bleaching and to survive in conditions that kill other corals. This is an encouraging finding.

What happens in polar regions?

Polar ecosystems are also very vulnerable to climate change. Their temperatures are increasing more rapidly than elsewhere (more than 5 times the global average). Warming ocean currents have been speeding up the melting of the Arctic sea ice sheet and the decline and breakup of Antarctic ice shelves. The Arctic ice sheets have been shrinking, with the lowest recorded level in the summer of 2012. It is predicted that the Arctic will be totally ice-free during the summer in less than 30 years. Greenland is losing about 100 billion tons of ice annually as a result of melting. Sea levels are now projected to rise much faster than predicted by the Intergovernmental Panel on Climate Change (IPCC) in 2007, because of this accelerated melting, which further threatens coastal habitats. As the temperature has risen, plankton blooms typical of the region have decreased, and the phytoplankton community has shifted from large species to smaller ones. This shift has affected the zooplankton. Shrimp-like krill, which are inefficient at grazing on small phytoplankton, are declining, while salps, which are efficient, are increasing. Krill also depend on diminishing sea ice for their reproduction. Furthermore, according to a study by Schofield and colleagues, other species that depend on ice, like Adelie penguins in Antarctica, are also decreasing, while other penguin species have increased. Changing wind patterns also affect Antarctica's plankton. Retreating sea ice and stronger winds have caused seawater to mix more deeply, a process that moves phytoplankton into deeper water, which has less light for photosynthesis. As a result, phytoplankton are declining, resulting in fewer krill (important food for baleen whales) and fish larvae. Krill are also affected by the loss of sea ice which is a refuge from predators.

The loss of sea ice also will stimulate major increases in shipping over the North Pole and Arctic Ocean in the future. Ships traveling in the Northwest Passage and through the central Arctic Ocean will likely bring new, potentially invasive, species to the Arctic as well as to northern ports. As the Arctic ice melts, new ports will be connected and shorter passages between existing ports will provide new opportunities for invasive species to spread. Shorter routes also mean that more organisms attached to the hull or in ballast water will survive the voyage. Invasive species, as a type of biological pollution, will be discussed at length in the next chapter.

Can climate change affect the distribution of species?

In response to warming, the geographical ranges of marine species are likely to change, including migration to higher latitudes (toward the poles) and to deeper depths where the temperature is more suitable. Phytoplankton are predicted to move toward the poles and away from the equator. If the oceans continue to warm as predicted, there will be a further decline in the abundance and diversity of phytoplankton in tropical waters and a shift toward the poles. The poleward movement of many marine animals has already been observed. However, animals that already live in polar regions are finding their habitat shrinking as the ice melts. Polar bears, for example, require sea ice, which is disappearing from the

Arctic at an alarming rate. The breeding population of chinstrap penguins has declined significantly as Antarctic temperatures have warmed. Two of the three chief penguin species in the Antarctic Peninsula—chinstrap and Adélie—are declining in a region where the temperatures over the last 60 years have warmed by 3°C (5°F). In contrast, Gentoo penguins are expanding both in numbers and in range.

Fish can respond to changes in ocean temperature by moving poleward to avoid warmer temperatures, or moving into deeper water. As water warms, fishermen are finding some new species that come from warmer regions. A 2009 report by the National Oceanic and Atmospheric Administration's Northeast Fisheries Science Center found that about half of the species it studied were shifting their range further north or into deeper colder water, including Atlantic cod, haddock, and hake—keystones of New England's ground fishery. The commercial lobster fishery is disappearing in southern New England. If animals cannot change their geographic or depth distribution, there may be changes in growth, reproduction, and mortality rates. Warmer water may lead to loss of productivity, but also to the opening of new fishing opportunities, depending on interactions between climate impacts, fishing grounds and fishing fleets.

Can climate change have effects on aquaculture?

Aquaculture is the fastest growing food sector in the world, according to the UN Food and Agriculture Organization (FAO), with most of the production coming from the developing world, where it makes a major contribution to the economy. Currently about half the world's seafood comes from aquaculture, and the proportion is expected to grow. Traditional fisheries are thought to be near their maximum capacity and future increases in seafood production will need to come largely from aquaculture. Animals can grow faster in warmer water provided they have enough food, which could be a boon for aquaculture and fish native to warmer waters could be farmed in new places. However, fish and shellfish disease is a greater problem for aquaculture in tropical countries. Diseases are more deadly and progress quicker in warmer climates. Outbreaks in tropical regions can wipe out entire fish stocks in a relatively short time, with major consequences for the economy and food security. Such outbreaks could become more severe with climate change.

Can climate change affect the size of animals?

As the climate changes, many species are expected to shift to smaller sizes. One reason for this is the need for oxygen. Aquatic animals are sensitive to low oxygen, which would likely accompany climate change. A recent study tested how organisms' mass changed with temperature. With each 1° C increase in temperature, aquatic animals that were 100 mg reduced their body mass by 5%, while land animals of the same size reduced their mass by only 0.5%. Using computer modeling, scientists found that fish sizes could shrink by about 20% from 2000 to 2050, due to warmer temperatures and less oxygen. There has already been a decline in growth and body size of North Atlantic cod in the United States, Canada, and Europe in response to warmer water. Smaller fish can have economic consequences on communities that depend on fish for food and trade.

Can climate change affect predator/prey interactions?

Temperature stress can affect predator/prey interactions. Many rocky shore intertidal organisms already live very close to their thermal tolerance limits. At cooler sites, mussels and barnacles are able to live high on up the shore, above the range of their aquatic predators (mainly sea stars). However, as temperatures rise they are forced to live lower down, placing them at the same level as predatory sea stars. Daily high temperatures during the summer months at sites in California have increased by almost 3.5° C (6.3° F) in the last 60 years, causing the upper limits of the habitats to retreat 50 cm (about 20 inches) down the shore, while the location of predators and the position of the lower limit have remained constant. Additional effects on predator/prey interactions come from ocean acidification, to be discussed later.

What effects can happen from sea level rise?

Sea level rise (SLR) is caused by thermal expansion of the warmer ocean water and by melting glaciers and ice sheets that contribute new water to the ocean. Although average global sea level remained relatively constant for almost 3,000 years, it increased by about 17 cm (7 inches) during the twentieth century, and is projected to rise by 40–80 cm by 2100. Over the twentieth century, global sea level increased at an average rate of about 2 mm per year, substantially greater than the rate of the previous three millennia. Measurements from 1993 to 2008 indicate that sea level is already rising twice as fast as in previous decades and is already exceeding the rise predicted by climate models. There are also differences in the amount of SLR in different parts of the earth. Although there is considerable variability associated with these and other estimates, 25 to 50% of SLR since 1960 has been attributed to thermal expansion. Small glaciers and ice caps shrunk considerably during the twentieth century and freshwater runoff from melting land-based ice will increase in the future. However, over the past 20 years melting ice sheets have become the biggest contributors to SLR, and will remain the dominant contributor in the twenty-first century if current trends continue. Sea level rise could be up to 1 m by 2100.

Studies indicate that we have already committed ourselves to a SLR of 1.1 m (3.6 ft) by the year 3000 as a result of greenhouse gas emissions up to now. This could be more severe, depending on the how much mitigation will take place. If we were to follow the high emissions scenario of the IPCC, a sea-level rise of 6.8 m could be expected in the next thousand years. The two other IPCC scenarios project SLRs of 2.1 and 4.1 meters. Rising sea levels could make entire areas, even island nations, uninhabitable or extremely vulnerable to flooding and storms. Because of dense concentrations of humans and development in coastal zones, many countries are vulnerable to SLR and coastal flooding. Tens of millions of people around the world are already exposed to coastal flooding from tropical cyclones. Global warming has the potential to increase flooding from more severe hurricanes and sea level rise. Developing countries, particularly small islands and low-lying areas, are especially vulnerable and have limited capacity to adapt to rising sea levels or to recover. Low-lying areas in developed countries such as Long Island, New York and South Florida in the United States are also at great risk. Coastal populations are particularly vulnerable to natural disasters including tsunamis, floods, and hurricanes. Since over a third of the world's population lives in coastal zones within 100 kilometers (62 miles) of the shore, the effects could be disastrous. According to IPCC, many millions more people will be flooded due to SLR by the 2080s.

SLR affects natural intertidal ecological communities such as salt marshes and mangroves at the edge of the water. These communities will have to migrate inland or increase their elevation in order to avoid being submerged by rising seas. As these are important habitats for birds and marine animals that use them as nursery habitats, many species are at risk if these wetlands cannot either migrate inland or increase their elevation. In many areas, marshes are not expected to be able to increase their elevation fast enough to keep up with SLR. However, if storms transport new sediments into the marshes, they may be able to increase elevation and persist for a longer time. In developed areas there are roads, houses, and other man-made structures just landward of the marshes, which prevent them from migrating inland.

Why is sea level rising faster than was predicted?

The IPCC report in 2007 projected a global SLR between 0.2–0.5 m by the year 2100. Current measurements meet or exceed the high end of that range, however, and suggest a rise of 1 m (3.3 feet) or more by the end of the century. The reason for the underestimate is that the models did not include critical feedbacks that speed everything up. These feedbacks are melting of Arctic sea ice and the Greenland ice cap. While ice is bright and reflects much of the sun's radiation back (this is called albedo), water is dark and absorbs it, causing more warming. Melting sea ice—which is already in the ocean—does not itself raise the sea level, but when it melts it releases more freshwater from the Arctic, which is then replaced by inflows of saltier warmer water from the south. That warmer water pushes the Arctic toward more ice-free waters, which, because of their dark color, absorb sunlight rather than reflect it back into space the way ice does. The more open water there is, the more heat is trapped in the water, and the warmer things can get. There are gigantic stores of ice in Greenland and Antarctica that are melting. This was clear in the summer of 2012 when Greenland had a record-setting melt. Another missing feedback is the groundwater being extracted all over the world to mitigate droughts. That water is ultimately added to the oceans. All these feedbacks will speed up SLR.

What can be done about sea level rise?

A major challenge is how to both mitigate and adapt to the impacts of climate change since impacts are now inevitable even if aggressive action is taken quickly to reduce greenhouse gas emissions, which seems unlikely. In order to protect coastal cities and towns, adaptation involves improving and increasing salt marshes and mangroves that reduce storm surges, and building sea walls and other structures to hold back the ocean. Restoring and constructing coastal wetlands, oyster reefs, and coral reefs is one approach to reducing the amount of surge during storms, which are predicted to become more severe in the future. Vegetation and reefs can reduce current speeds and dissipate some of the energy because these organisms provide drag force in the water. Another approach is to retreat—to move houses and communities back from the shore, a strategy that runs into considerable political opposition. Perhaps if it were called something else—perhaps "move-back"—it would not sound like defeat and cause so much opposition.

What is pH?

The pH is a scale from 0 to 14 used to measure of hydrogen ion concentration in water. A pH of 7 is neutral, and represents equal amounts of hydrogen ion (H^+) and hydroxide ion (OH^-). Pure water is neutral. Solutions with a pH less than 7 have more H^+ than OH^- and are acidic, while solutions with a pH above 7 are basic or alkaline; 0 is as strong as an acid can be, and 14 is the strongest alkali. Seawater is slightly alkaline, with a pH around 8.2. Since pH is a logarithmic scale, a difference of one pH unit is equivalent to a tenfold difference in hydrogen ion concentration. So a pH decrease from 8.2 to 8.1 represents a 30% increase in acidity (even though the water is not really acidic, but is less alkaline).

What is ocean acidification?

When CO_2 from fossil fuel burning enters the atmosphere, about 1/3 of it ends up dissolving in the ocean. While this is good for us since it slows down global warming, it is bad for the ocean. In the ocean, the CO_2 combines with water to form carbonic acid, which becomes bicarbonate ions (HCO_3^-) and hydrogen ions (H^+), reducing the pH of the water by making it more acidic. Since the industrial age began, the pH of the oceans has declined by 0.1 pH unit, which, because the scale is logarithmic, represents a 30% increase in acidity. This is lower than the pH has been in 20 million years. The extent to which human activities have raised ocean acidity has been difficult to calculate because it varies naturally between seasons, from one year to the next, and among regions and specific locations. In addition, direct observations go back only 30 years. If CO_2 emissions continue at the present rate, often called the business-as-usual scenario, models project an average worldwide decrease of 0.2–0.3 units by 2100 on top of what has already happened, doubling the current acidity. The Southern Ocean is an important carbon sink—about 40% of the CO_2 absorbed by the oceans enters there. Rather than being absorbed uniformly into the deep ocean in vast areas, CO_2 is drawn down by currents. Winds, currents, and massive whirlpools (eddies) that carry warm and cold water around the ocean create localized pathways and acidic patches.

Reduced pH or ocean acidification (OA) threatens not only the ecological health of the oceans, but also the economic well-being of the people and industries that depend on a healthy productive marine environment. Eutrophication—algal blooms resulting from increased nutrients (see Chapter 2)—is another source of CO_2 in coastal waters. When combined with CO_2 from the atmosphere, the release of CO_2 from decaying organic matter is speeding up the acidification of coastal seawater. The pH in the lower part of the Chesapeake Bay is declining at a rate three times faster than the open ocean, partly because of nutrients from farming and other activities. These combined effects make the job of minimizing harmful impacts of OA that much more difficult.

What effects are produced by ocean acidification?

The increased acidity of the oceans is expected to harm a wide range of ocean life—particularly those with calcium-containing shells (Figure 9.2). Many organisms use calcium and carbonate ions from seawater to produce calcium carbonate for shells.

Decreased pH reduces the saturation of calcium carbonate, making it more difficult for some organisms to accumulate calcium and carbonate to make their hard shells and skeletons. Two common mineral forms of calcium carbonate are aragonite and calcite. Those animals that use aragonite (corals, pteropods, and bivalves) are expected to be more severely affected than calcite calcifiers (coralline algae, sea urchins) because of differences in solubility—aragonite is a more soluble mineral form than calcite. (It is interesting to note that otoliths, the bony structures in fish ears, appear to get larger in acidified conditions rather than smaller as would be predicted.) It appears that larval mollusks and some other calcifying organisms are already experiencing harmful effects on shell formation at some locations. Delicate corals may face an even greater risk because they require very high levels of carbonate to build their skeletons. Acidity slows reef-building, which could lower the resiliency of corals and lead to erosion. Since coral reefs are home to a host of other organisms, their loss would have extensive effects throughout the marine

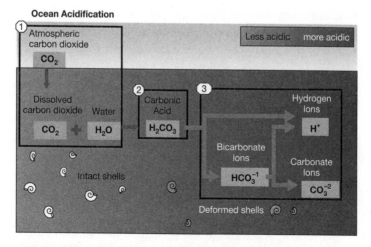

Figure 9.2 Ocean Acidification (taken from Chesapeake Quarterly, Maryland Sea Grant College Program)

environment and have profound social impacts in the tropics—especially on fishing and tourism. The loss of coral reefs would also reduce the protection that they offer coastal communities against storm surges and hurricanes.

Hasn't ocean pH changed in the past? Why is this different? Will marine organisms be able to adapt?

Ocean pH has changed in past geological ages. But the rate of change then was very slow—over many thousands of years. Today's pH change is extremely fast—over one or two hundred years. In the slow changes, processes like rock weathering and seafloor mineral dissolution could counteract some of the changes. But now the change is happening too fast for slow geological processes to counteract. While some marine organisms will be able to tolerate these conditions or evolve adaptations, the changes may be happening too fast for many organisms to tolerate or be able to evolve adaptations.

Which species are most threatened by ocean acidification?

Ocean acidification impairs the process of calcification for building shells, which means that calcareous plankton (including some phytoplankton at the base of oceanic food webs), corals, shellfish—anything that builds a shell—is at risk. Among the plankton in the oceans are tiny mollusks called pteropods that play an important role in the oceanic food web. Because they produce an aragonite shell, they are expected to be very sensitive to ocean acidification. They make up a large part of the diet of Alaska's juvenile pink salmon, which could be affected indirectly through loss of food. Shellfish with weaker thinner shells would be less able to resist shell-crushing predators.

Some acidity is natural in some regions. Water off the Pacific coast of the United States already has a low carbonate saturation state. When surface winds blow the top layer of water out from coastal regions, deeper water with high acidity (called corrosive water) can upwell, and harm shellfish. Periodic upwelling of CO_2-rich water has already happened on the US West Coast, where larval oyster survival has been very low for several years because of these events. A few decades ago, such upwelling events weren't as acidic and probably wouldn't have been cause for concern. Hatcheries are having trouble producing and rearing larval oysters. There has been a reduced natural set of juvenile oysters in some estuaries where the commercial shellfish industry relies on natural reproduction of oysters. This is due largely to the upwelling of acidic deeper water. Workers at Oregon's Whiskey Creek Shellfish Hatchery suspected that low pH water was killing their oyster larvae. Working with Oregon State University and NOAA, they were able to show that that was the case, and now they monitor the pH of the ocean and time their water intakes to ensure that oysters are exposed to less acidic water. A small investment in pH-monitoring equipment saved the industry millions of dollars.

In addition to the stress of warmer water, corals are very sensitive to acidification and construct weaker shells. To add insult to injury, when seawater is both acidic and warm (as will be the case), corals become even more fragile due to microborers such as algae, blue-green algae, and fungi that bore tiny holes in the coral, further weakening the skeleton. Corals will not only have less material to build their reefs, but older parts will erode faster due to both acidity and boring. If current trends continue, there will be major decreases in global coral reefs with declines in associated fishes and invertebrates.

Increasing CO_2 may be an additional stress driving a shift from corals to seaweeds on reefs. In an experiment, coral deaths from contact with a reef seaweed increased two- to threefold between background CO_2 levels and a level of CO_2 projected for late twenty-first century. Thus, coral reefs may become more susceptible to overgrowth by seaweeds and be replaced by them. Other members of the reef community, however, may enable corals to be more resilient. Herbivorous fishes and invertebrates such as sea urchins play an important role in reef health by mowing down and eating weedy algae and thus clearing settling spots for young corals. Those herbivores may help damaged reefs to recover. However, not all species of corals respond the same way. Some species have a degree of tolerance to lower pH, while others experience harmful carryover effects through developmental stages or even generations after short-term exposure. Bleaching, acidification, and diseases are expected to compound each other, and will reduce survival, growth, reproduction, larval development, settlement, and postsettlement development. Interactions with local stresses such as pollution, sedimentation, and overfishing will likely intensify the effects of climate change.

Ocean acidification effects are not restricted to shell production. Mussels use stiff, stretchy structures called byssus threads to attach onto surfaces. In lower pH, the threads break more easily and lose elasticity; the mussel's ability to hang on drops by about 40% in more acidic water. Effects have been seen on behavior and development of a number of marine animals. Fish use gills to regulate pH balance, but the early larval stages don't have gills and cannot regulate pH balance in this way. Exposure of eggs and larvae of a common estuarine fish to elevated CO_2 severely reduced survival and growth. The eggs were more vulnerable to high CO_2-induced mortality than the larvae. Atlantic longfin squid eggs raised in seawater with elevated CO_2 were slower to hatch than those raised in normal seawater. Mineral structures called statoliths, which help the squid sense movement, were smaller in acidified water; they had more pores and were oddly-shaped. With abnormal statoliths, the squid might have trouble orienting and swimming.

Behavior is also altered in many animals. For example, young clownfish (familiar as the cartoon fish, Nemo) normally stay close to the reef in which they live. But as the water becomes more acidic they tend to wander farther and farther from home. This boldness is not good for their survival because the farther away they go, the more likely they are to get eaten by predators. Furthermore, Philip Munday and colleagues studying coral reefs next to natural seeps near New Guinea (where carbon dioxide bubbles are made by volcanic activity), have found that fish lose their fear of predators. Living in this acidic environment makes small reef fish become attracted to the smell of their potential predators. In another example of reduced olfactory behavior, homing ability is impaired in cardinalfish in low pH. All these modified behaviors can increase mortality. But predatory behavior can also be impaired. The brown dottyback (a coral reef fish) in elevated CO_2 levels shifted their behavior from preference to avoidance of the smell of injured prey, and decreased their feeding activity.

The Chilean abalone, a snail that adheres to rocks along wave-swept shores, quickly rights and reattaches itself when it is dislodged, an important skill. But when CO_2 levels were increased (pH decreased), snails were slow to right themselves or did not do so at all. Responses of hermit crabs to food odors were disrupted under reduced pH. Crabs in reduced pH seawater had lower antennular flicking rates (the sniffing response), were less successful in locating an odor source, and had reduced activity compared to those in untreated seawater.

Since there is great variation in sensitivity to OA, we can expect that some species will thrive while others will diminish greatly, thus causing major changes in marine communities. For example, some large crustaceans such as crabs and lobsters do not seem to be impaired by excess CO_2, but instead seem to grow larger, despite their need for calcium in their shells. Under high CO_2, they have been found to molt faster. After molting, they undergo a growth spurt while in the soft-shell stage. Extra carbon speeds the molt cycle so that they become bigger, potentially less vulnerable to predators and possibly better predators themselves. These studies are all preliminary, and much work needs to be done to understand effects in a variety of organisms. In acidified conditions mollusks grow less and have weaker shells, making them more vulnerable to predators. Sea urchins also seem resistant to OA, largely because they have genes that provide resistance and can evolve rapidly. Researchers raised larvae in water with either low or high CO_2, sampled the larvae, and used DNA-sequencing tools to determine which parts of the genetic makeup changed. By looking at the function of each gene that changed, researchers were able to identify which particular genes were critical for sea urchin survival under acidic conditions.

Ocean acidification can degrade entire ecosystems, resulting in homogenized communities dominated by fewer plants and animals. In the waters by Castello Aragonese, an island off the coast of Italy, volcanic vents naturally release CO_2, creating different levels of acidity, which provide a glimpse of what the future communities could look like. Three zones—low, high, and extremely high acidity—representing conditions of the present day, 2100, and 2500 respectively were selected for sampling. Researchers removed animals and vegetation from rocks and examined the rocks periodically for recovery. In more acidic water the number and variety of species was reduced. In both high and extremely high acidic plots, fleshy algae increased and took over, because sea urchins and other grazers were either not present or did not graze on the algae, while they did so in the lower acidity zone. Calcareous grazers are important in maintaining the balance in marine ecosystems and are among the most vulnerable species to OA.

How can organisms protect themselves against effects of warming and acidification?

Many studies consist of placing marine animals in laboratory tanks with low pH water for a few days or months to see how they respond. Fish and shellfish larvae often fail to thrive and don't grow as big or live as long as those in more alkaline waters. But some species show substantial resilience. Unlike most laboratories, organisms in the real ocean live in a community with many different species and a complex web of interactions. Some species are competitors for space and food; others are potential prey or predators. Limited laboratory studies also cannot tell you the long-term effects or if a species can adapt to acidification. Our present understanding relies mostly on results from short-term studies. Longer studies may reveal that some species can adapt over time. Animals can be impaired when abruptly exposed to elevated CO_2, but individuals that are gradually acclimated to high CO_2 may be able to adjust over the long term. For example, coral in high CO_2 for one week reduced calcification by about 25% in a pH decrease of 0.1 units. In contrast, the coral could acclimate to this pH over six months, and have even a slightly greater calcification rate. This shows acclimation of a coral to ocean acidification. Some mollusks have also been able to increase their tolerance to low pH through acclimation. In one study, elevated CO_2 caused oyster larvae to reduce growth, slow their development, and reduce survival. But when adult oysters were exposed to elevated CO_2 while their gonads were ripening, the larvae they later produced were larger and developed faster in the high CO_2 conditions. Also, selectively bred larvae were more resistant to elevated CO_2 than wild larvae. Thus, some marine organisms may be able to acclimate or adapt to elevated CO_2.

Another approach is studying wild populations that have already adapted to acidic waters, which occur naturally in some parts of the world. For example, along North America's West Coast, the waters off Oregon have low pH due to upwelling. While this does severe harm to oysters as described above, sea urchin larvae there tolerate acidic water better than ones from California. This suggests that these sea urchins have the resilience and genetic variation sufficient to tolerate ocean acidification. Researchers have found indications that some corals have genes that provide increased resilience to acidification.

Some corals can adapt to higher temperatures and resist bleaching. For example, corals in the Persian Gulf withstand summer temperatures up to 10°C higher than corals elsewhere and have recovered from extreme temperature events in 10 years or less. Their heat-tolerance suggests that other coral populations might also be able to adapt to increasing temperatures. Recovery from bleaching events is faster when there are enough herbivorous fishes on the reef to keep algae in check. Researchers found that because of overfishing, densities of herbivores on Red Sea reefs near Saudi Arabia were far lower than other Red Sea reefs, which can explain the very slow recovery of these reefs. Some corals can adapt to acidification. Scientists have found that corals near the island of Palau are able to thrive in highly acidic water. So it is possible that other corals in other areas will be capable of adapting to acidification as well.

What economic effects could result from Ocean Acidification?

Ocean acidification may impact tourism and fisheries and the jobs and revenue that depend on them. Regions that depend heavily on coral reef tourism or coral reef based fisheries will have severe impacts from OA, which could decrease revenue if the quality of reefs or fish harvests decline. It is also important to realize that changes in shellfish harvests, coral reef-associated industries, or tourism will affect other businesses and communities that depend on the affected industry. This could really amplify the overall economic effects of ocean acidification. Commercial fisheries, especially for shellfish, can be severely affected. Vulnerability of different countries to decreases in mollusk harvests will depend on their nutritional and economic dependence on mollusks. Countries with low adaptability, high dependence on mollusks, and rapidly growing populations will be the most vulnerable.

What can we do to mitigate effects of ocean acidification?

Reduction of CO_2 emissions and increased sequestering of carbon are approaches that are being considered. While there is much discussion about planting more trees and restoring rain forests, seagrasses and mangroves fix carbon at a much higher rate than land-based systems and could be an approach to climate mitigation (known as "blue carbon"), which also preserves the important ecosystem services of these habitats. Because these approaches are long-term, expensive, and unlikely to be enacted soon, Washington State has decided to try to buy time for itself. They will monitor ocean acidity carefully and create an acidity budget—an assessment of how much acidity is coming from which sources. Then it will attempt to reduce carbon inputs from land-based sources such as agricultural and urban runoff. They will also develop practical steps to offset carbon, like planting seagrasses. They will include an extensive campaign to educate the public, business leaders, and policymakers about the risks of increasing OA.

There are many efforts underway internationally to restore and plant new salt marshes and seagrasses not only as buffers for climate change, but for the habitat they provide for marine animals and the shore protection they provide to human communities. There are also efforts underway to restore damaged coral reefs. These projects are very labor intensive, involving growing small pieces of coral in the laboratory and then cementing them in place on a reef. Restoring oyster reefs has become very popular in the United States for a number of reasons. The oyster reefs provide habitat for a wide variety of other animals, the reefs serve as a buffer against storm surges, and their calcium-containing shells can help to counteract decreasing pH. In addition, oysters filter huge amounts of water and can help to combat eutrophication by eating a lot of phytoplankton.

Geoengineering technologies—technical fixes—can play a role in tackling climate change. Iron fertilization is one that has actually been tried in order to increase CO_2 uptake from the atmosphere into the ocean. In areas where the growth of phytoplankton is limited by low availability of iron, extra iron is added. This stimulates plant growth, increasing photosynthesis and hence carbon uptake into the ecosystem. Some of this extra carbon should end up in the deep ocean, carried there in the bodies of dead organisms. Twelve large-scale experiments have been undertaken, mainly in the Southern Ocean, with mixed results. Overall, they have produced little evidence that the technique will reliably sequester carbon. Most scientists think that such efforts are not likely to be effective. New calculations take into account not only the CO_2 that will be sequestered in the deep sea but also subtract losses due to ventilation, greenhouse gas production, and the burning of fossil fuels in order to produce the iron salts, transport them, and distribute them at sea. These calculations suggest that a single iron fertilization event will sequester just 10 tons of carbon/sq km at a cost of almost $500/ton of CO_2. In July 2012, Russ George of Planktos, Inc. dropped 100 tons of iron sulfate into the Pacific Ocean off Canada's West Coast. He claimed that this study was meant to mitigate climate change by spurring the growth of plankton. Satellites show that there was a plankton bloom over 10,000 square kilometers. The Canadian government launched an investigation to determine whether he violated international treaties including the Convention on Biological Diversity and the London Convention on Dumping of Wastes at Sea. Much of the scientific community condemned his actions because the project had violated international agreements, including a moratorium declared by the United Nations Convention on Biological Diversity.

In contrast, energy efficiency is a win-win situation. Much more must be done to develop energy efficient cars, buildings, and appliances. Efficient technologies can contribute large emission reductions, since they offer high cost savings and can significantly reduce emissions. Unfortunately, small-scale innovations that improve efficiency go unnoticed because they don't have the glamour of solar panels and wind turbines, and don't benefit from the market interests and political influence.

Recycling can combat climate change because it reduces the need to mine and process new materials which produces greenhouse gases. By reducing the amount of trash, we cut both CO_2 emissions from incinerating waste and methane (a potent greenhouse gas) emissions from trash decomposing in landfills. Recycling paper reduces the cutting down of trees that absorb greenhouse gases. The US recycling rate has been rising steadily since the 1970s when the rate was only 7%; 34% of trash was recycled in 2010. This is good, but we still have a way to go. The almost 10 million tons of paper containers and packaging we throw away annually could save the energy equivalent of 1.6 billion gallons of gasoline.

Despite all these efforts, global CO_2 continues to rise. In the United States, the Energy Information Administration (EIA) statistics for total domestic carbon dioxide emissions reveal a 2.32 % increase in US carbon emissions during 2013, over 2012 levels. It is urgent that remedies to halt ocean degradation be established; the rate, speed, and negative impacts of climate change in the global ocean are greater and faster than previously thought.

10

BIOLOGICAL POLLUTION

Where does microbial pollution come from?

Microbial pollution comes from sewage that has not been properly treated. Fecal contamination is a concern because some of the microbes it contains can cause illness. High levels of pathogens may be present in effluent from sewage plants that do not have secondary treatment. Older cities that have combined sewer systems combining storm sewers and household and industrial waste can discharge untreated sewage during heavy rains when the volume of water exceeds the capacity of the system. There is no national record-keeping of how many illnesses are caused by sewage releases, but it is estimated that as many as 20 million people each year become ill from drinking water containing pathogens from untreated waste that entered the water upstream from drinking water sources. Nonpoint runoff is another source of microbes into coastal waters—from animal feces, livestock operations, or dense concentrations of wild animals. Hepatitis A virus, and pathogenic bacteria (e.g., *Salmonella, Listeria monocytogenes, Vibrio cholerae,* and *Vibrio parahaemolyticus*) have been reported in coastal waters.

Microbiological contamination can build up in marine life (shellfish) when sewage is released to coastal waters or arrives in river flow. Bacteria and viruses from humans and animals, usually attached to fine particulate matter, can affect bathing water quality, another potential source of illness. A study estimated that as many as four million people become sick each year in California from swimming in water with pollution linked to sewage. In urban rivers such as the Hudson near New York City, bacteria have been found that are resistant to antibiotics. The stretches of the river with the most sewage-indicator bacteria also generally contained the most antibiotic-resistant ones. The resistant bacteria include potentially pathogenic strains of *Pseudomonas, Acinetobacter, Proteus,* and *Escherichia* (as in *E. coli*). Scientists have also isolated *Vibrio* and other bacteria from seawater and sand of recreation beaches on the Baltic Sea and found that they were resistant to many antibiotics. Overall, however, the microbiological quality of coastal waters has been improving in recent decades following better wastewater treatment.

How is microbial pollution detected?

Enterococci (e.g., *E. coli*) are microbes that are naturally occurring in the digestive systems of mammals and birds, but are also opportunistic pathogens that cause millions of infections annually. Because they are shed in human and animal feces, can be easily cultured in the laboratory, and their density can predict health risks from exposure to polluted recreational waters, they are used as surrogates for waterborne pathogens and as fecal indicator bacteria in water testing laboratories around the world.

What kind of diseases may result from exposure?

Microbial pollution by pathogens from sewage or animal waste is a major concern for drinking water supplies, but is also an issue in coastal waters where swimmers may become ill after rain has washed bacteria in from combined sewers or in runoff. In comparison with drinking water, infections and illness due to recreational water contact are generally mild and difficult to detect. Even when illness is severe, it is often difficult to attribute a particular case to coastal water pollution.

Epidemiological studies have shown gastrointestinal and respiratory infections associated with recreational water, which is why bathing beaches may be closed after major rain storms when elevated bacterial pollution has been detected. Another concern is accumulation of pathogenic bacteria and viruses in edible shellfish, which can cause more severe illness in people who eat them. Gastroenteritis and hepatitis A are the most important diseases transmitted through shellfish; however, cholera and typhoid fever (from *Vibrio cholerae* and *Salmonella typhi*) were the first to be linked to eating contaminated shellfish. Viral outbreaks are also associated with eating contaminated shellfish. Initially the analysis of disease outbreaks was based on epidemiological data, but advances in molecular biology and the ability to detect low levels of enteric viruses in shellfish now provide better information on shellfish as a source of disease. Officials close shellfish beds when tests indicate elevated bacterial levels. *Vibrio parahaemolyticus* is a leading cause of seafood-borne bacterial gastroenteritis. An outbreak of *V. parahaemolyticus*-related gastroenteritis was linked to consumption of raw oysters in Washington State.

Humans are not the only ones at risk from pathogens washed into the water. Many sea otter deaths have been attributed to diseases known in terrestrial mammals. M. A. Miller and colleagues have found an association between sea otter deaths from a particular pathogen and increased stream flow that took place 30 to 60 days earlier, supporting the idea that runoff from land brought fecal pathogens from land animals to the sea otters.

How can people know if it is safe to swim at their favorite beach?

In 2000 the US Congress passed the Beach Act, which extended pollution protections to coastal waters and required states to set up monitoring programs for pathogens. Municipalities and counties generally close beaches when the counts exceed certain standards. However, most health departments monitor for bacteria once a week, which may not be often enough. Models can be used to predict how long after a rainfall a particular beach will be affected by microbial pollution, but if you are concerned, it might be best not to swim for a couple of days after a major rainfall. The Waterkeeper Alliance has developed a swim guide and a website at www.theswimguide.org, which provides up-to-date information about many beaches.

What are invasive species?

When a species arrives in a new environment, it is unlikely to have the natural controls that kept its population numbers in balance in its native area. Without control by predators, parasites, or disease, some species increase rapidly, to the point where they can take over their new environment and harm native species. Marine organisms have been moved around the world for thousands of years by ocean currents and attached to driftwood, and more recently aided by human activities. What is new is the speed and volume at which marine organisms are now transported. Recently, marine invasive species have had major impacts on biodiversity, ecosystems, fisheries, human health, and economics. Although most introductions fail, a small percentage of species can thrive and become a problem in the new area. Certain traits—for example tolerance of environmental stress—characterize species that become invasive. Environments that are stressed (e.g., polluted, eutrophic, low in native diversity) appear to be more vulnerable to invasions. As demonstrated by Andrew Cohen and James Carlton, San Francisco Bay appears to be a hotspot for invasions, containing many of the invasive marine species in western North America.

How do they get to new locations?

Aquatic species can be transported by various means (vectors), either accidentally or on purpose: in ship ballast water or by attaching to hulls, as hitchhikers clinging to boots or scuba gear, as consignments of live organisms traded to provide live bait or food, and as symbionts or parasites carried by other organisms. The mechanisms, extent, and consequences of marine and estuarine invasions have been reviewed by Gregory Ruiz and colleagues. In the 1800s, trans-Atlantic shipping increased dramatically and many species were transported between Europe and the East Coast of North America. The periwinkle snail (*Littorina littorea*) arrived in the early 1800s and is now widespread on rocky shores from Canada to New Jersey. It has greatly altered the ecology of these shores. Increased trade and shipping moves more organisms around the world in ship ballast water in one month, than used to be moved in a century. It is estimated that 7,000 species are carried around the world in ballast water every day and 10 billion tons of ballast water are moved each year. Which species will survive in a new location, and which ones may thrive and cause problems cannnot easily be predicted, a phenomenon that James Carlton has termed "ecological roulette."

Fouling by attached organisms is another important vector. Organisms attach to the hull, to propellers and propeller shafts, anchors, and anchor chains. Paints with tributyltin are being phased out due to environmental concerns. With less effective antifoulants, fouling will likely transport more organisms in the future.

Many marine species including oysters, marsh grasses, and fish were deliberately introduced for food or for erosion control, with little knowledge of the impacts they could have. Fish and shellfish have been intentionally introduced all over the world for aquaculture, providing food and jobs, but they can escape and become a threat to native species, ecosystem function, or livelihoods. Pathogens or parasites associated with the species that are moved can infect native species and even humans. Atlantic salmon are reared in ocean net-pens in Washington State and British Columbia. Many escape each year, and they have been recovered in both saltwater and freshwater in Washington State, British Columbia, and Alaska.

Recreational fishing can also spread invasive species. Bait worms from Maine are popular throughout the country. They are commonly packed in seaweed which contains many other organisms. If the seaweed is discarded, it or the organisms on it can colonize new areas. Wading boots, recreational boats, and trailers can pick up organisms at one location and move them elsewhere.

The aquarium trade can also be responsible for marine invasions. Many people keep exotic fish, marine plants, invertebrates, or corals in aquariums. One of the most infamous marine invaders, a strain of the tropical seaweed *Caulerpa taxifolia*, now carpets large areas of the Mediterranean Sea. Molecular studies by O. Jousson and colleagues indicated that it was derived from an aquarium strain. Since the alga was first spotted right under windows of the Oceanographic Museum in Monaco, it probably came from that aquarium. The lionfish, one of the most devastating invaders in the Caribbean, probably originated from aquarium pets that were released.

Debris washed out to sea by the Japanese tsunami in 2011 is washing ashore in North America, carrying with it large numbers of hitchhikers; thus far over 60 Japanese species have come on floating debris to the west coast of the United States and Canada. Of special concern are docks, piers, buoys, and vessels that were in seawater at the time of the tsunami and would have already had populations of attached organisms. What makes this different from boat transport is that boats move too quickly between ports for many organisms to hang on. Also, the communities transported on slow-moving tsunami debris can arrive along the whole coastline rather than just at ports. Of great concern is the Northern Pacific seastar, a shallow-water species that eats shellfish. After it invaded Australia its population grew to 12 million in two years and it had major impacts on aquaculture. A fast-growing seaweed called wakame kelp has also been found on much of the Japanese debris. Shellfish including blue mussels, Pacific oysters, brown barnacles, and clams from Japan have also been discovered, along with worms and sea urchins. In 2012 a 66 ft-long commercial shipping dock washed ashore in Oregon. Of the 100 species attached to its sides, two-thirds were native to Asia, including seaweeds, mussels, sea stars, barnacles, crabs, and oysters, which survived at sea for 14 months and about 5,000 miles. Fish native to East Asia were discovered on a Japanese fishing boat set adrift by the tsunami that washed ashore in Washington state two years later. Sea anemones, scallops, crabs, worms, and sea cucumbers were also found on the boat. The West Coast states and Hawaii have developed response plans. A Japan Tsunami Marine Debris Taxonomic Assessment Team, with experts familiar with marine organisms of the North Pacific will examine photographs quickly to indicate if a species is potentially invasive so that decision-makers can determine a response strategy.

What are some invasive marine fishes and what harm do they do?

Lionfish (*Pterois volitans*), native to the Indo-Pacific and available in the tropical fish trade, were spotted first in the early 1990s off the coast of Florida and were believed to have either been released from aquaria or when Hurricane Andrew flooded aquarium and pet stores near the coast (Figure 10.1). In the Atlantic they are taking food and habitat from native fishes that are important to the local ecology and economy. They have no natural predators, and are now found in large numbers in nearly all marine habitats in the Atlantic along the Southeast United States and continuing along the South American coast, as well as in the Gulf of Mexico and Caribbean, to which they have spread. They have a potent venom in their spines that deters predators. Their sting normally is not deadly, but it is extremely painful. As shown by Mark Albins and other investigators, their densities have surpassed some native reef fish in many areas, and they grow larger and are far more abundant in the invaded areas than they are in their Pacific native range;

Figure 10.1 Lionfish *Pterois volitans* (photo from NOAA)

in some areas they make up almost half of the total biomass of predators. The ecological impacts of this invasion range from disrupting the structure and function of reef communities to impacts on commercial fishing and tourism. Lionfish eat ecologically important species such as algae eaters (e.g., parrotfish) that keep algae in check on coral reefs. On heavily invaded reefs, lionfish can remove over 60% of prey fish, some of which include economically important species like snapper and grouper. The socioeconomic impacts can be severe, especially to fishing and tourism, which are critically important to many Caribbean and Atlantic countries. Off the coast of North Carolina they are eating so well that, like obese people, they are found to have globs of fatty tissue on their internal organs—not a normal condition for a fish.

What are some invasive jellies and what harm do they do?

One of the worst marine invasions occurred in the early 1980s when the North American comb jelly *Mnemiopsis leidyi* (Figure

Figure 10.2 Comb jelly *Mnemiopsis leidyi* (photo from NOAA)

10.2) arrived in the Black Sea. Comb jellies, or ctenophores, superficially resemble jellyfish but are biologically quite different. They do not sting, and belong to a different phylum (Ctenophora), so are not really jellyfish. In its native Atlantic estuaries, abundance is restricted by predators and parasites, and it tolerates a wide range of temperature and salinity. It reaches 10 cm in length and eats zooplankton, including fish eggs and larvae. Populations can reach very high densities. When it arrived, it rapidly took hold in the Black Sea. By 1989, there were about a billion tons of them eating vast quantities of fish eggs and larvae as well as the zooplankton that commercially important fish feed on, leading to the collapse of fish stocks and the ecosystem of the Black Sea, as documented by T. Shiganova. Genetic analyses showed that they had come from both the Gulf of Mexico (e.g., Florida) and the northern part of the native range (e.g., Rhode Island). The high genetic diversity in the Black Sea population indicates release of a large number and multiple invasions, which is consistent with ballast water transport and their extensive distribution in the Atlantic. In a strange turn of events, in 1997, another comb jelly,

Beroe ovata, invaded the Black Sea. A larger species, it feeds on *M. leidyi* and caused a dramatic fall in their number, helping the ecosystem to recover. Growing populations of zooplankton, phytoplankton, and fish have been seen. It is possible to use *Beroe ovata* as a biological control for *Mnemiopsis leidyi*. However, purposely using another alien species for control of an invader should be a last resort, given that it carries its own risks of becoming invasive too.

What are some invasive crabs and what harm do they do?

The green crab (Figure 10.3) (*Carcinus maenas*) is native to the Atlantic coasts of Europe and Northern Africa, where it lives on protected rocky shores, pebbly beaches, mud flats, and tidal marshes. It thrives in a wide range of salinity and temperature, and has invaded South Africa, Australia, and both coasts of North America. Its larvae spend about two months in the plankton, dispersing many miles along the coast. Then they are swept by tides and currents into coastal waters and

Figure 10.3 Green crab *Carcinus maenas* (photo from Peddrick Weis)

estuaries, where they molt and settle out as juveniles. If conditions are suitable they will survive and reproduce, establishing a new population. Green crabs arrived on the Atlantic Coast of the United States in the 1800s, probably on ship hulls, and settled in coastal bays from New Jersey to Cape Cod. Later they began moving north, and their arrival in Maine coincided with dramatic declines in the soft clam fishery. They are a major predator of soft-shelled clams and quahogs. They also feed on oysters, worms, and small crustaceans. They can crack open clams and mussels faster than other crabs, and can out-compete native crabs for food. A second major invasion was detected in 1989 in San Francisco Bay, where they probably arrived as larvae in ballast water or in seagrass or kelp used in packing shipments of lobsters and bait worms to the West Coast. Ted Grosholz and Gregory Ruiz documented their spread and effects. Their arrival was associated with losses of up to 50% of the Manila clams in California. As they continue to move north there is concern for Dungeness crab, oyster, and clam fisheries in the Pacific Northwest. They also are detrimental to eelgrass beds since adults uproot the eelgrass, and juveniles graze on it.

In North America green crabs have fewer parasites and actually grow larger than they do back home in European waters, which may contribute to their success. On the East Coast, snails and mussels that have been living with green crabs for over a century have developed thicker shells as a defense, making them harder to crush than those that have not been exposed to green crabs. The crabs, in turn, develop stronger claws—an example of an evolutionary "arms race."

The Chinese mitten crab (*Erocheir sinensis*) is a burrowing crab native to the Yellow Sea in Korea and China. It gets its name from the dense patches of hairs on its claws. They are believed to have been accidentally released in ballast water in the early 1900s in Germany. In the 1920s and 1930s they expanded into many northern European rivers and estuaries. The Thames River in England has also had a population explosion. They travel long distances upstream into freshwater, feeding on native species. They also burrow into stream and river banks leading to bank collapse. British zoologists fear that this crab could both eat and out-compete vulnerable freshwater species and that native crayfish (which are already in decline) could be affected. Economic impacts in Europe from fishery losses due to mitten crabs are estimated at around 80 million Euros. Many animals prey on them, including raccoons, river otters, wading birds, and fishes, but they apparently do not eat enough of the crabs to slow down their invasion.

Considered a delicacy in Asia, live mitten crabs have been imported illegally into seafood stores in the United States. They became established on the West Coast in the 1990s and are considered a threat to native invertebrates, to the structure of freshwater and estuarine communities, and to some commercial fisheries. They may imperil California's endangered salmon, due to their appetite for salmon eggs. They can walk on land and enter new rivers to disperse far and wide. Another problem in California is their impact on water diversion and on fish salvage facilities. In the summer of 2006 they appeared in Chesapeake Bay, and by 2007 were spotted in Delaware Bay and the New York/New Jersey vicinity. New York and New Jersey have issued alerts for crabbers to report any sightings. Sightings along the East Coast have been sporadic without any indication (so far) that a population is established and growing.

The Asian Shore Crab, *Hemigrapsus sanguineus*, a small species, was first observed by John McDermott in New Jersey in 1988 after it probably arrived as larvae in ballast water. It has extended its range to Maine and North Carolina, becoming abundant in pebbly intertidal and shallow water habitats. They reproduce readily in a wide range of conditions, and are found in very high densities; in some areas they have displaced green crabs, possibly because they prey on small green crabs. *H. sanguineus* is now the dominant crab in rocky intertidal habitats along much of the northeast coast of the United States. There is little evidence that they have major predators or parasites. They also like the moist, shady environment created by cord grass and mussels in salt marshes. The cord grass attracts ribbed mussels by giving them something to attach to; the mussels, in turn, give the crabs crevices in which to avoid predators, a process referred to as a facilitation cascade. The cord grass provides valuable shade for both mussels and crabs. In this case, the crabs' use of the habitat does not seem to crowd out native species. It is encouraging that the salt marsh habitat can apparently accommodate this new resident without severe problems. However, its broad ecosystem effects and economic impacts are as yet unclear, and there are indications that their populations are declining in favor of native species.

What are some invasive sedentary attached organisms and what harm do they do?

Tunicates or sea squirts are fouling organisms that attach to hard substrates. Invasive tunicates are found mostly associated with artificial structures like floating docks, pilings, and aquaculture installations, but they also settle on natural habitats. Some invasive tunicates, (golden star tunicate, *Botryllus schlosseri*; violet tunicate, *Botrylloides violaceus*) settle on eel grass blades and reduce light penetration, thereby reducing the growth and survival of the grasses, which are important habitats for numerous animals. The sea squirt, *Didemnum vexillum*, has a history of invading and overgrowing marine communities in temperate waters, including New England and mid-Atlantic coasts, as summarized by Gretchen Lambert. It reproduces rapidly, spreads easily, and can colonize and dominate large areas of benthic habitat. They can overgrow native organisms such as mussel beds. Areas with large amounts of open space, regardless of species richness, are vulnerable to *Didemnum*. Processes that fragment its colonies may accelerate the spread of this invader, which is able to reproduce while in a fragmented state. Thus, trying to remove it by cleaning off fouled surfaces and dredging probably aids the spread of this species unless it can be contained and totally removed from the water.

What are some invasive seaweeds and what harm do they do?

Large areas of seabed in the northern Mediterranean are now carpeted by *Caulerpa taxifolia*, an invasive seaweed that pushes out native marine life, disrupts ecosystems, and affects fishermen's livelihoods. In the late twentieth century it was very popular in the aquarium trade. The public aquarium in Monaco apparently released small amounts of the seaweed into the wild. After remaining as a patch in front of the aquarium for a while, it expanded and covered the seabed along 190 km of coast. By 2001, it had spread to many other harbors around the Mediterranean on boat anchors or fishing nets. It starts out by overgrowing and shading native seaweeds or seagrasses, and then affects animals that rely on the native species for food. Animals that cannot move away quickly, such as shellfish, are smothered. This seaweed protects itself by producing a toxin, so there are relatively few species that can eat it. One species that does is the Mediterranean bream, which accumulates the toxins but is not directly harmed.

Japanese kelp or wakame (*Undaria pinnatifidia*) is native to Japan, China, and Korea, where it is harvested for food. It tolerates a wide range of conditions and can grow on any hard surface, including rope, boat hulls, bottles, mollusk shells, and other seaweeds. It may form dense forests outcompeting native species for space and light. It was intentionally introduced into France for commercial use and then spread to the United Kingdom, Spain, and Argentina. It was unintentionally introduced into Australia, New Zealand, and Italy. It can interfere with aquaculture by attaching to cages or ropes, either slowing the growth of or displacing the farmed species.

What are some invasive marsh plants and what harm do they do?

The common reed *Phragmites australis* (Figure 10.4) is native to the United States. However, a different genetic strain from Europe is a major invader in East Coast brackish and freshwater marshes. It outcompetes native plants, replacing diverse plant assemblages in freshwater and brackish wetlands. In brackish marshes its effects on aquatic animals vary; some species are not affected by its takeover of the marsh, while killifish (mummichogs) (and people) clearly prefer the native cordgrass, *Spartina alterniflora*. Detritus from *Phragmites* gets into food webs the same as the detritus from *Spartina*, so it has a similar trophic function. The invader can sequester pollutants better than the native cordgrass, and is an effective buffer against storm surge, so it has both negative and positive effects.

The Atlantic cordgrass *Spartina alterniflora*, native to the East Coast of the United States, is a highly valued native. It has been planted elsewhere for coastal protection and sand dune stabilization. However, when moved to new areas, such as the

Figure 10.4 Common reed *Phragmites australis* (photo from Peddrick Weis)

West Coast, it can become invasive, as documented by Curtis Daehler and Donald Strong of the University of California, Davis. Its invasion of Willapa Bay, Washington is transforming vast areas of tidal mudflats into dense vegetation, affecting migratory waterfowl, shorebirds, and wading birds that forage in the open mudflats. In San Francisco Bay, it has hybridized with native *Spartina* species, threatening the native flora in marsh areas. The hybrids are tougher than their parent species and become even better invaders. Cordgrass was introduced to China in 1979 from the United States for reducing coastal erosion. It grows vigorously in China and has spread over much of the coastline, where it is competing with native *Phragmites* (some irony here?). It has also invaded mangrove areas in which the canopy was opened by human disturbance. Some fear that it could gradually replace these mangroves in midsalinity regions of Chinese estuaries.

Can an alien species do some good?

Some alien species have been found to benefit native species and the local environment. In Chesapeake Bay, for example, the exotic red alga *Gracilaria vermiculophylla* has flourished and dispersed widely. It turns out to provide nursery habitat for juvenile blue crabs at places where native eelgrass has declined (largely due to eutrophication). *Phragmites* is an effective barrier providing storm and flood protection and is very effective at sequestering pollutants. Furthermore, it produces more detritus and litter than *Spartina,* so it increases the marsh elevation and may enable tidal marshes to keep ahead of rising sea levels.

In degraded New England salt marshes, green crabs can do some good. Because of overfishing, natural predators of the marsh crab, *Sesarma reticulatum,* have been depleted so that marsh crab populations exploded. This native crab eats large amounts of cordgrass, so their dense populations destroyed large areas of salt marsh. Mark Bertness and Tyler Coverdale of Brown University found that when green crabs invaded these marshes they ate and displaced the *Sesarma,* and actually promoted recovery of the marsh grass.

What can be done to prevent new invasive species from arriving?

It is much harder to eradicate an alien species in a marine environment than on land, but not impossible. If eradication is not possible some type of control may be achievable, even though it will need to be ongoing and is very expensive. In all cases, it is better to prevent the introduction in the first place. Surveillance and monitoring are essential to detecting a new arrival in time to deal with it before it turns into a problem.

Prevention through management of ballast water is gaining much attention, since ballast water discharges are a major vector of introductions. Because of international regulations, vessels are now supposed to exchange ballast water in the open ocean before arriving at their destination. Organisms in ballast water taken on in a port are likely to be adapted to estuarine or river conditions so they will not survive in the open ocean when released. The ship refills its tanks with ocean water; oceanic species should not be able to survive when released in ports and harbors. However, it is not always possible for ships to use this method because of safety concerns, such as in rough seas. Also, emptying the water does not remove the sediments on the bottom of the ballast tanks. Many organisms inhabit these sediments, including dinoflagellate cysts, some of which are toxic, presenting a possibility of harmful algal bloom (HAB) problems in the new location. Also living on the bottom of ballast tanks are adult invertebrates including green crabs, mud crabs, periwinkles, soft shell clams, worms, and blue mussels, which are not removed by midocean water exchange and live their whole adult lives cruising on the seas. While their densities are usually low, invasion risk may still be significant, especially during reproductive seasons. Gravid female crabs may carry thousands to millions of eggs, and after release of larvae new clutches may be fertilized by sperm retained from an earlier mating. While ballast water exchange can help to reduce marine invasions, it should not be the only measure. Research is ongoing into methods of destroying organisms in ballast water using sterilization, ozone, or heat treatment. Another option is to build treatment plants in ports that take ballast water from the ships and sterilize it before releasing it or returning it to another ship. Implementation of onshore treatment should be practical in busy ports that receive high volumes of ballast water. This also would not take care of the invertebrates cruising around in the mud on the bottom of the ballast tanks.

To reduce the risk of new invaders, standards are being proposed that establish upper concentration limits for organisms in ballast discharge. Standards have been established by the International Maritime Organization (IMO), the US Coast Guard, the US Environmental Protection Agency, and individual states in the United States. Australia, Canada, and New Zealand have established ballast water regulations. The IMO established discharge standards based on the number of viable organisms per volume of ballast discharge for different organism size classes. The EPA has developed more stringent numeric standards limiting the release of organisms in ballast water.

Education for aquarists, fishers, and others can go a long way toward reducing the introduction and spread of new invasive species. In some areas new techniques are being used in the aquaculture industry to reduce the risk of invasion. For example, farmed mussels can be manipulated to have additional sets of chromosomes, making them sterile and thereby reducing the risk of wild populations establishing.

New technology can improve monitoring and control. The International Union for the Conservation of Nature (IUCN) Centre for Mediterranean Cooperation has released a new app for smart phones to help managers of marine protected areas control the spread of invasive species in the Mediterranean Sea. The app, which includes an identification guide to the most important invasive species found in the Mediterranean, will help to spot invasive species in marine protected areas so that monitoring and control programs can be put in place quickly before they damage native species.

What can be done after an invasive species has arrived?

Once a species has arrived in a new area it is important to locate it and take action very quickly, before it has a chance to establish and spread. This can be difficult. Eradication successes are rare in the marine environment. But if an invader is found while it occupies a relatively small area, it may be eradicated if the response is quick enough. Surveys are important and can be site-specific, such as in ports where species may be introduced, or species-specific (targeting those that are known to pose high risks). Surveys can be carried out by organizations with responsibility for detecting invasive species.

An eradication of the brown mussel *Perna perna* from a deep soft-sediment area in New Zealand was undertaken following its discovery among fouling organisms physically removed from a drilling rig. A total of 227 dredge tows were undertaken, removing about 35 tons of material that had been removed from the rig, and which was disposed of in a landfill. The removal of these bivalves from relatively deep (>40 m) soft sediments indicates that, with appropriate tools and other resources (including substantial funding), eradication is feasible even in challenging environments. They did not claim to have removed all the mussels, but said that when total elimination is not feasible, density-based success criteria can be developed that can effectively mitigate risks.

A relatively small *Caulerpa* infestation in Southern California was spotted and quickly eradicated by covering the seaweed with plastic sheets and poisoning it with chlorine. The cost of this eradication was $2.33 million in 2000–2001, for control and monitoring, with an ongoing annual surveillance cost of $1.2 million until 2004. Application of coarse sea salt was used with moderate success in Australia to eradicate *Caulerpa taxifolia*. Croatia attempted eradication by covering the seaweed with plastic sheeting, which was reasonably successful in a limited area. Eradication has also occurred in South Australia and New South Wales, and manual removal by scuba divers was successful in eradicating a small patch in the French Mediterranean. However, these methods are very resource intensive and if even a tiny piece is missed, the species can easily recover.

Caulerpa sold in the aquarium trade has the potential to invade US waters. Surveys of southern California aquarium retail stores in 2000–2001 showed that 26 of 50 stores sold at least one *Caulerpa* species, with seven stores selling *C. taxifolia*. In late 2001, California banned the importation, sale, or possession of nine *Caulerpa* species. To determine the effectiveness of the ban, *Caulerpa* availability at previously sampled stores in Southern California was investigated four years after the ban. Of 43 stores, 23 still sold *Caulerpa* with four stores selling *C. taxifolia*, suggesting that the ban has not been very effective and that the aquarium trade is still a source for distributing *Caulerpa*.

What can be done after a species has become abundant?

It is extremely difficult to control a marine organism once it is established. Many marsh restoration projects in the East Coast of the United States involve the removal of *Phragmites* and replanting of *Spartina*, which is very labor intensive and expensive and may require numerous applications of toxic herbicides. This type of restoration may also involve lowering the marsh elevation to favor the growth of cordgrass. However, such restored marshes may eventually become inundated sooner by rising sea levels. In some cases, re-establishing normal tidal flow is sufficient for the *Phragmites* to decline and

Spartina to return because the reed cannot tolerate the higher salinity while cordgrass can.

A successful eradication of the black striped mussels from marinas in Australia took place. Chemicals were used to kill everything in the marinas, including all native marine life. The operation involved chemically treating three marinas and 420 vessels, engaging 270 people (including sharpshooters to protect divers from crocodiles) over four weeks at a total cost of 2.2 million Australian dollars. In Hawaii's Kaneohe Bay a "supersucker" vacuum device is used to remove alien algae *(Gracilaria salicornia)* that forms a thick mat smothering and killing coral.

Because lionfish are so damaging ecologically and environmentally, countries have developed outreach and management strategies to reduce their populations. Control efforts can reduce lionfish densities and impacts, and in stressed systems even a small reduction in impact could have long-term benefits. Removal can be by professionals or staff of nongovernmental organizations (NGOs), using divers and volunteers. Fishing and spearfishing tournaments have been organized that include education about basic biology, ecology, impacts, collecting, and handling techniques. Data collected during lionfish derbies can be used to monitor populations over space and time. Some derbies in Mexico and the Bahamas have removed over 2,000 lionfish in a single day. Monthly ongoing removal contests offer prizes and recognition to divers and dive operators, some of whom include lionfish collecting in their regular dive activities; customers are encouraged to search for lionfish during their dives and notify the divemaster. Lionfish are rather sedentary and bold, allowing a close approach with a spear, but they have quick escape reflexes. In Bermuda, divers are allowed to remove lionfish from areas traditionally off-limits. They must attend a workshop for safety. The fish can also be removed effectively with nets, and they go into fish traps. A study followed divers who removed lionfish weekly from several sites off Little Cayman Island,

Caribbean. Researchers asked divers not to remove lionfish from one particular area so it could be used as a control site. At removal sites lionfish density decreased, and those that remained were smaller, while fish numbers increased at the control site. When the study began, many lionfish were about 16 inches long, but at the end of the study the removed fish were less than half this length. Size is important, as larger fish consume bigger prey and lay a lot more eggs. One potential problem that may result from intensive culling is that the fish can learn to change their behavior to become less visible to divers. A recent study compared behavior of lionfish from an area where spear fishing had occurred with fish from areas where hunting had not taken place. On culled reefs, fewer fish were active during the day, and they hid themselves much more effectively.

Divers in some areas have attempted to entice top predators (sharks, barracudas, groupers, snappers) to consume captured lionfish, in the hope that predators will learn to hunt and prey on them. To date, there is no evidence that predators are learning to prey on lionfish. Unexpected effects of these activities include aggressive behavior of predators during encounters with divers, with subsequent injuries to the divers.

Can invasive species be controlled by eating them?

If you can't beat 'em, eat 'em: if you can't persuade sharks to eat the lionfish, you can encourage people to do so.

Lionfish, it turns out, are very tasty, and many countries are developing incentives for people to eat them. Many fishers rely on their catch to provide much of the diet for their families and consider lionfish good eating. Eating the catch is also a strong incentive for recreational fishers. Promoting consumption of lionfish encourages their removal, but education in handling the fish, with their toxic spines, is essential. Restaurants in Mexico, the Caymans, and Caribbean islands are beginning to serve them, and cookbooks have been written. The news magazine of North Carolina Sea Grant, *Coastwatch*, published several lionfish recipes in their autumn 2013 issue. There has been considerable interest in developing commercial markets for lionfish. Consistent, high-volume removal efforts may halt population growth, diminishing their devastating effects on native fish populations. In Belize's Barrier Reef Reserve System, efforts are underway to do just that. Conservationists from Blue Ventures, a British nongovernmental organization, are working together with local communities and the Placencia Producers' Cooperative Society in developing a new international export market for the fish.

One criticism of fishing is the potential for juveniles to be left in favor of larger, market-ready fish. One way to prevent this is to also create a market for juveniles in the aquarium trade. Puerto Rico currently exports approximately 200 to 300 juvenile lionfish per week to supply the US aquarium trade, and Florida Keys collectors also remove and sell juveniles. While the aquarium trade is likely the initial source of the invasion, the number of fish released was few. Even if a small percentage of collected fish were to end up back in the wild, the numbers removed are greater and could contribute to a reduction in impacts. While there are economic benefits to fishers and collectors for lionfish removals, there is concern that allowing aquarium trade in invasive lionfish could lead to additional introductions.

In South East Asia mitten crabs are a delicacy, especially the ovaries and testes. Consequently, they are eaten primarily during the fall migration period when the gonads are ripening. Crabs in Europe are being considered for consumption. The Dutch have developed a fishery for them. A large population is established in the Thames River, England, and commercial exploitation is being considered. A report considers that the population is large enough to support artisanal fishing and that fishing should be limited to certain months. This could reduce crab numbers and provide financial benefits for local fishermen. Although the Thames crabs had accumulated some contaminants, the report concluded that the harvesting of mitten crabs from the Thames for moderate consumption would not pose serious risks.

On the other hand, one fishy invader that would never do for eating is a species of pufferfish that has invaded Lebanon, Cyprus, Turkey, Greece, and Egypt. This Indo-Pacific native eats large amounts of smaller fish and damages fishing nets. It arrived in the Mediterranean about ten years ago and contains a highly potent toxin in its tissues.

Even though we may be able to control invasive species by harvesting them as food, some caution must be taken. A possible problem is that developing a market creates pressure to maintain the species and if it becomes an economic resource, people may try to extend the market in new areas by introducing it. Turning an invasive species into an economic resource may prompt the local community to protect the species. Therefore, projects aimed at controlling invasive species through human consumption should be carefully examined for both benefits and potential problems.

11

REGULATING AND REDUCING POLLUTION

What is the Ocean Health Index?

Using a new comprehensive index designed to assess the benefits to people of healthy oceans, a group of scientists led by Ben Halpern have evaluated the ecological, social, economic, and political conditions for every coastal country in the world. The Ocean Health Index is the first broad, quantitative assessment of the important relationships between people and the ocean in terms of the benefits we derive from the ocean. Instead of simply assuming any human presence is negative, it evaluates how our impacts affect the things we care about. The Ocean Health Index defines a healthy ocean as one that delivers a range of benefits to people both now and in the future. A healthy ocean can maintain or increase its benefits to us (food and services) in the long term, without jeopardizing the marine life that provides these benefits. This index aggregates ten different measures into a score of how well the oceans are doing. It is not a measure of how pristine the ocean is, but measures how sustainably it is providing the things we care about. The measures—which include water quality and factors such as food provision, carbon storage, tourism value, fisheries, aquaculture, coastal protection, and biodiversity—were chosen to reflect both ecosystem sustainability and human needs. The index was derived from existing data on such things as the percentage of live coral on reefs, to the percentage of coastal people served by adequate sanitary facilities, and the extent of arctic sea ice. The developers of the Index analyzed over 200 data sets and measured each country's score against reference points that set standards of maximum sustainable use. The Index can provide guidance for ocean policy since it includes current status, trends, and factors affecting sustainability. It should enable managers to focus on important actions to improve the health of the ocean, promote awareness of the state of the oceans, and be a guide for decision-makers. The overall global score was 60 out of 100 (barely passing). Almost one-third of the world's countries earned a score of 50 or lower. But 5% of the nations scored higher than 70, showing that there are successes. Developing nations, which are generally less able to plan and control ocean usage, tended to have lower scores, and developed nations generally had higher scores. The United States received a score of 63; Britain, 61; India, 52; and China, 51. Jarvis Island (an uninhabited 1¼ square mile island in the Pacific) had the highest score of 86.

The Clean Water category averaged pollution intensity from chemicals, excessive nutrients, pathogens, and trash; the target (for a score of 100) was to have zero pollution. Russia was rated first for clean water, with a score of 97, and Benin in West Africa was the lowest, with a score of 22. Four measures contributed to the score: the amount of pollution, the trends (percent rise or fall over the past five years), the stresses, and resilience (actions taken to reduce stresses). Countries with long coastlines relative to their size (e.g., islands) tended to have higher scores. Some small islands got high scores because they are relatively uninhabited (e.g. Jarvis Island) or located in remote regions with reduced pressures. On the other hand, Russia and Canada, which have very long coastlines, scored well because they have long uninhabited coastlines along the Arctic, and because they manage water quality well in their populated regions.

What is the Law of the Sea?

The United Nations Convention on the Territorial Sea and the Contiguous Zone (1958) outlined the rights and responsibilities of States in their offshore waters. In 1982, the UN Convention on the Law of the Sea (LOS) further developed the role of States in their marine areas and beyond. While the United States ratified the 1958 Convention, by the end of 2013, it had not signed onto the 1982 Convention. The LOS sets forth a legal framework for the sea, the seabed, and its subsoil, plus the protection of the marine environment and its resources. It requires countries to adopt regulations and laws to control marine pollution and establishes jurisdictional limits on the ocean area that countries may claim, including a 12-mile territorial sea limit and a 200-mile exclusive economic zone limit. (The practice of nations claiming jurisdictional rights over activities in waters off their coast dates back to the seventeenth century, where a three-nautical mile territorial sea was recognized as the limit of a state's control over activities off its coast.) The United States recognizes that the principles in the 1982 LOS Convention reflect customary international law, but it is not bound by the agreement itself.

What is MARPOL?

In 2008 the United States became a Party to Annex VI of the International Convention for the Prevention of Pollution from Ships (MARPOL). MARPOL Annex VI addresses vessel air pollution; large, diesel-powered ocean-going vessels such as container ships; tankers; cruise ships; and bulk carriers, which must limit their emissions of nitrogen oxides and use cleaner-burning fuels to reduce sulfur dioxide (SO_2) emissions. Parties may also designate areas off their coasts—called SO_2 emission control areas—where more stringent controls apply. Since air pollution comes down in precipitation to become water pollution, this will improve the marine environment as well as air quality. Other types of marine pollution are also covered in MARPOL, as described in Chapter 1.

What is the London Convention?

The London Convention and London Protocol establish global rules and standards for reducing and controlling pollution of the marine environment from dumping. The main objective of the London Convention is to prevent indiscriminate disposal at sea of wastes that could create hazards to human health or marine life, damage amenities, or interfere with other legitimate uses of the sea. The 1982 United Nations Convention on the Law of the Sea (LOS) directs states to adopt laws and regulations that are no less effective than the rules and standards of the London Convention and Protocol. It follows a black list/ grey list approach to ocean dumping: Annex I materials (black list) generally may not be dumped (though certain materials may be allowed if present only as trace contaminants or rapidly rendered harmless), and Annex II materials (grey list) require special care. The United States implements the Convention's requirements through the Marine Protection, Research, and Sanctuaries Act.

What national laws in the United States promote clean water?

The Clean Water Act (CWA) is the primary federal law governing water pollution. Passed in 1972, the act set goals of eliminating releases of high amounts of toxic substances into water, eliminating additional water pollution by 1985, and ensuring that surface waters would meet standards necessary for human sports and recreation (fishable and swimmable) by 1983. (These goals were not met by 1983 and have not been met yet.) The CWA uses two methods to protect water quality: monitoring the water quality, and controlling discharges from point sources. The National Pollutant Discharge Elimination System (NPDES) is a permit system used by the

Environmental Protection Agency (EPA) for regulating point sources such as sewage plants or industrial facilities like manufacturing, mining, and oil and gas extraction. Point sources may not discharge pollutants to surface waters without a permit from EPA, in partnership with state environmental agencies. The permit describes what and how much is allowed to be discharged. EPA requires technology-based standards, which are developed for each category of dischargers based on pollution control technologies, without regard to the conditions of a particular water body. The idea was to establish a basic standard for all facilities in a category, using the best-available technology. Water bodies that do not meet water quality standards with technology-based controls alone are placed on a list of water bodies not meeting standards. If water quality still remains impaired (probably because many sources discharge into the same waterbody), then the permit agency (state or EPA) may add water quality-based limitations to the permit. These limitations are more stringent and require the facility to upgrade and install additional controls. They must develop a total maximum daily load (TMDL), which is a calculation of the maximum amount of a pollutant that a water body can receive and still meet water quality standards. The TMDL is determined after study of the water body and the pollutant sources contributing to the noncompliance. Water quality standards include designated uses (the best being fishable and swimmable), water quality criteria, and antidegradation policy. Water quality criteria can be numeric levels of specific pollutants. A narrative criterion serves as the basis to limit toxicity of waste to aquatic species. A biological criterion is based on the aquatic community—the number and types of species present in a water body. A nutrient criterion protects against nutrient overenrichment, and a sediment criterion describes conditions of contaminated and uncontaminated sediments in order to avoid undesirable effects.

After passage of the Clean Water Act, communities upgraded their sewage treatment plants, factories upgraded their technology, and the water quality in most areas of the country improved considerably. Nonpoint sources (e.g., runoff) are not regulated to the extent that point sources are, and remain a major continuing source of impairment of water bodies even though the CWA nonpoint source program provides grants for demonstration projects, technology transfer, education, training, assistance, and related activities to reduce nonpoint source pollution. It would appear that this voluntary approach is not as effective as regulations that can be enforced.

The Marine Protection, Research, and Sanctuaries Act (MPRSA) or Ocean Dumping Act was passed by Congress in 1972. The MPRSA regulates the transportation of waste for ocean dumping beyond the territorial limit (three miles from shore) and prevents or strictly limits dumping material that "would adversely affect human health, welfare, or amenities, or the marine environment, ecological systems, or economic potentialities." A government report indicated that in 1968, 38 million tons of dredged material (which was 34% polluted), 4.5 million tons of industrial wastes, 4.5 million tons of sewage sludge (which was significantly contaminated with metals), and 0.5 million tons of construction and demolition debris were dumped in US waters. The MPRSA authorized EPA to regulate ocean dumping of materials including industrial waste, sewage sludge, biological agents, radioactive agents, garbage, chemicals, and other wastes, into the territorial waters of the United States through a permit program. The EPA can issue permits for dumping of materials other than dredge spoils if the agency determines, after a full public process, that the discharge will not unreasonably degrade or endanger human health or welfare or the marine environment. Although ocean dumping was rare, it was a major controversy in the New York/New Jersey area for many years. The 12 Mile Dump Site, 12 miles off the coast of New Jersey, was where all of New York City's sewage sludge as well as sludge from some New Jersey communities was dumped—over six million tons annually. The area became contaminated with high levels of organic contaminants, pathogens, and metals, which formed a layer of black muck on the bottom where few organisms could live. In 1987, after over 60 years of operation and amid public protest, the dump site was moved to deeper water 106 miles from shore, where the wastes could disperse more effectively, but the controversy continued and a federal law was passed banning all ocean dumping. On June 29, 1992, the *New York Times* reported: "Late this afternoon the ocean barge Spring Brook will slip quietly into the East River and head to sea, carrying for the last time one of America's least loved cargoes: 400 tons of New York City's processed sewage. It has been four years since Congress voted to ban the common practice of using the ocean as a municipal chamber pot, and with the Federal deadline set for tomorrow, New York is the only city that still does it. For environmentalists and many politicians, the final barge journey will be a moment of triumph, one they say will make the planet a cleaner, healthier place."

Superfund is the federal government's program to clean up hazardous waste sites. It is also enforced by the EPA. Officially known as the Comprehensive Environmental Response, Compensation, and Liability Act of 1980 (CERCLA), it was enacted after the discovery of toxic waste dumps in places such as Love Canal and Times Beach in the 1970s. It allows the EPA to clean up such sites and to compel responsible parties to either perform cleanups or reimburse the government for doing so. The process is complicated and includes assessment of sites, placing them onto a national priorities list, and establishing and implementing cleanup plans. This is a long-term process, and some sites have waited over 30 years to be cleaned up. Part of the reason is that establishing liability for the contamination and determining who is responsible for paying for the cleanup leads to lengthy lawsuits. There are often many responsible parties, some of which may be companies that have since gone out of business or have been swallowed up by other companies (which are still considered responsible for cleaning up the mess). While most Superfund sites are on land, many coastal areas have been polluted by industries that discharged their toxic wastes into tidal marshes or estuaries. A number of estuarine Superfund sites in New York and New Jersey discussed in this book have been cleaned up (e.g., the cadmium-contaminated Foundry Cove in the Hudson River), are in the process of being cleaned up (polychlorinated biphenyls (PCBs) in the Upper Hudson River; dioxin in the Passaic River), or are still awaiting cleanup (mercury-contaminated Berry's Creek in the Hackensack Meadowlands).

The Clean Air Act has also provided incidental benefits to coastal water quality. For example, air pollution controls have greatly reduced the amount of nitrogen that enters bodies of water. It had been estimated in the 1980s that about one-third of the nitrogen coming into Chesapeake Bay came from atmospheric sources. These sources were greatly reduced in the 1990s due to tailpipe emission standards for vehicles, changes in gasoline formulations, cleaner engines, and less pollution from ships and other entities that burn oil, gas, or coal. The 1990 Clean Air Act amendments aimed at curbing acid rain forced reductions in emissions of nitrogen oxides by utilities. While these were enacted to protect human health, they have had positive effects on water quality.

The presence of toxic chemicals in our waters is determined in part by laws that regulate and control which chemicals are in use. For chemicals used as pesticides, the law is FIFRA (the Federal Insecticide, Fungicide and Rodenticide Act). As discussed earlier, new pesticides must undergo some toxicity tests and toxicity is balanced against the benefits they provide. While some of the most notorious pesticides have been banned under FIFRA, others that are banned in Europe remain in use in the United States. In the case of tributyltin (TBT), Congress had so little faith in FIFRA's process that it passed a separate law restricting its use in antifouling paints on boats. For chemicals that are not pesticides, the federal law is TSCA, (the Toxic Substances Control Act). When this law was passed in the 1970s it "grandfathered" in tens of thousands of chemicals (estimated 60,000) already in use. For new chemicals, TSCA, unlike FIFRA, does not require any toxicity testing unless EPA puts forth a strong case that it ought to be tested. In other words, a chemical is innocent until proven guilty. As a result, in the four decades since the law was passed, only a handful of chemicals (out of an estimated 80,000 by now) have been restricted or banned. If the law took a more precautionary approach, our environment wouldn't be full of polybrominated diphenyl ethers and other harmful chemicals. Many people feel EPA should be empowered to demand more health and safety information from the chemical industry and shift the burden of proof to chemical companies. However, attempts to put teeth into this law and make its approach more conservative are strongly opposed by politicians who, interestingly, use that term to describe themselves.

What is NOAA's Role in the United States?

The National Oceanic and Atmospheric Administration plays an important role in administering the Law of the Sea: Its nautical charts provide the baseline that marks the inner limit of the territorial sea and the outer limit of internal waters, such as bays and rivers, which determines where US territorial waters begin for purposes of international law. The boundaries of offshore areas are revised due to shoreline changes such as accretion and erosion. The location of maritime boundaries can have far-reaching effects. NOAA works with other federal agencies, such as the US Department of State, to keep track of maritime boundaries and to represent such boundaries on navigational charts.

Although NOAA is a regulatory agency for fisheries, some parts of the Endangered Species Act, and the Coastal Zone Management Program, it does not regulate water pollution. However, it is involved in developing educational programs about marine pollution, especially nonpoint sources and marine debris, providing technical assistance and advice in the United

States and Caribbean countries to reduce pollution inputs, and in beach cleanups. In addition, the MPRSA charged the Secretary of Commerce (delegated to NOAA, within the Department of Commerce) with monitoring and researching ocean dumping and conducting research on long-term effects of pollution, over-fishing, and other human-induced changes in the ocean. NOAA has several laboratories that study marine pollution, including ocean acidification. NOAA scientists were heavily involved in oil spill research after the *Exxon Valdez* and Deepwater Horizon accidents. NOAA has had a Mussel Watch program for over three decades, analyzing bivalves from certain sites for contaminants. This long-term data has enabled in-depth knowledge over time about hot spots and seafood safety, and provides valuable data on background levels and trends of chemicals including about 50 polycyclic aromatic hydrocarbons (PAHs), which are potentially cancer-causing. When cleaning up an oil spill, for example, such long-term data provides important knowledge on how clean the place was before the spill.

How does the European Union regulate marine pollution?

The European Union (EU) has established a community framework for water protection and management. Member states must identify and analyze individual river basins and districts and then adopt management plans and measures for each body of water. The objectives include preventing and reducing pollution, promoting sustainable water use, environmental protection, improving aquatic ecosystems, and mitigating effects of floods and droughts. Its ultimate objective is to achieve good ecological and chemical status by 2015. Each member state is expected to analyze each river basin, review impacts of human activities, analyze the economics of water use, develop a list of areas requiring special protection, and survey all bodies of water used for extracting water for human consumption and producing more than 10 m^3 per day or serving more than 50 persons.

For marine waters, member states have to draw up strategies in cooperation with other countries to protect and restore marine ecosystems and to ensure the ecological sustainability of marine-related economic activities. Europe's seas are divided into four regions: the Baltic Sea, North-East Atlantic, Mediterranean, and Black Sea. The three seas are unique in that they are mostly enclosed. In each region, the member states must coordinate their actions with the other countries involved. At the regional level, member states must assess the ecological status of their waters and the impacts of human activities. This assessment covers the characteristics of these waters (physical and chemical features, types of habitat, animal and plant populations), an analysis of the main impacts and pressures from human activities (e.g., toxic contamination, eutrophication, nonindigenous species, damage by ship anchors), and an economic and social analysis of the use of these waters and cost of the degradation. Member states must then determine the "good ecological status" of the waters based on biodiversity, presence of nonindigenous species, stock health, the food chain, eutrophication, hydrographic conditions, contaminants, and noise pollution. On the basis of this evaluation, they must define objectives and indicators to achieve this status. Objectives must be measurable, consistent within a region, and tied to a definite timetable. Member states draw up specific measures to achieve the objectives, and must consider economic and social consequences. Before implementation, the measures are subject to impact assessments and cost/benefit analyses. Member states also establish coordinated monitoring programs to evaluate the status of the waters and the progress toward their objectives.

For regulating the manufacture and use of chemicals, a precautionary principle is used in the EU, a very different approach than that of the United States. Environmental quality standards are developed for priority substances or groups of substances that pose substantial risk. The Water Framework Directive establishes a list of 33 priority substances including metals, benzene, PAHs, and DDT. Regarding pollution from ships, legislation states that polluting discharges constitute in principle a criminal offence; this relates to discharges of oil or other noxious substances from vessels. In 2014 the European Parliament passed a directive aimed at cutting the use of thin single-use plastic bags by 50 percent by 2017. Many developing countries have legislation on the books but it is not always enforced.

What are some success stories?

Many highly contaminated waterways in the United States have been improving over the past few decades as a result of the Clean Water Act. The Elizabeth River in Virginia was highly degraded during its industrial heyday, with creosote wood preserving operations and ship repair yards. Toxic spills, explosions, and inadequate containment and disposal practices at these sites caused extensive sediment contamination from compounds such as polycyclic aromatic hydrocarbons (PAHs) and a variety of metals. Bacterial contamination, measured by fecal coliform levels, from upland runoff and discharges from malfunctioning sewage treatment plants, also contributed to the river's placement on the EPA's list of impaired waters. However, over the last few decades, the effects of long-term toxic discharges and sediment contamination have been reduced. There has been progress in improving the river's sediment and water quality. Vibrant salt marshes and productive oyster reefs can now be found along the river in the midst of industrial operations that now are partners in its restoration.

The Hackensack Meadowlands in Northern New Jersey has undergone an impressive recovery after decades of abuse. This twenty-one thousand acre marsh system has its eastern edge three miles west of Manhattan, and is the largest brackish marsh system in the New York/New Jersey area. As part of the most densely populated region of North America, it was disturbed by extensive development, drainage, diking, filling, garbage dumps, and sewage. Wetlands were considered useless and were polluted with industrial wastes, covered by asphalt, and used as legal and illegal waste dumps. Throughout the 1950s and 1960s, it was subject to uncontrolled dumping of millions of tons of garbage at 24 dumps covering 2,500 acres. Since no distinction was made between household waste and hazardous waste, all sorts of toxic materials including paint, petroleum, chemical, plastics, and pharmaceutical wastes were mixed with the garbage. This history has left hotspots of chromium, PCBs, mercury, and many other contaminants throughout the area. The river also received pig waste from rendering plants, and effluent from 13 sewage plants which, until the late 1960s, was mostly untreated. Most people considered the area an unpleasant wasteland. It has seven Superfund sites including Berry's Creek, one of the most mercury-contaminated sites in the nation. Just adjacent to the district is the lower Passaic River, one of the most dioxin-contaminated sites in the country.

The turnaround began in 1969 with the formation of the Hackensack Meadows Development Commission (HMDC) by an act of the New Jersey legislature to provide for the reclamation, planned development, and redevelopment of the area. The commission was responsible for waste management, development, and conservation, which are at best difficult to balance. They closed down and capped unregulated landfills, in some cases leaving toxic contaminants under a layer of dirt, so that leachate continues to ooze out into the river. They decreased illegal dumping, prohibited dumping of New York garbage, and cleaned up remaining landfills; all of these actions reduced the release of contaminants into the air and oozing into the river and wetlands. The federal Clean Water Act stimulated municipalities to upgrade and build effective sewage treatment plants that greatly reduced wastes coming into the water. Changes in the economic base also helped improve the environmental quality as polluting industries closed and nonpolluting businesses were established. As the water quality in the Hackensack River improved dramatically, there have been striking increases in the numbers of fish, bird, and invertebrate species. More than 50 species of fish now use the estuary for parts of their life cycles. Despite the fact that sediments remain contaminated and consumption of fish and crabs is prohibited, the lower Hackensack River has been declared an "essential fish habitat" by the National Marine Fisheries Service, an action mandated by Congress for each federally managed fish species. Much marsh restoration is ongoing, and the area is pleasant to visit. Social and recreational uses (e.g., ecotours) of the Meadowlands are increasing and provide benefits to urban populations, including awareness and appreciation of the environment and local wildlife. Recreational facilities such as parks, trails, overlooks, boardwalks, wildlife observation sites, an environmental park offering canoe trips, nature walks, bird watching, and an environmental center are functioning in the midst of this densely populated region just three miles from Manhattan.

How can we reduce pollution from aquaculture?

While floating cage cultures release fish waste, contaminants, and uneaten food, closed farms onshore contain their wastes and other byproducts, making them easier to handle. US fish farmers are experimenting with enclosed, recirculating systems, which filter wastewater and compost solid wastes to reduce impacts of untreated wastes. These farms can be located away from sensitive habitats where fish feed and breed. Freshwater tilapia, catfish, cobia, and trout are raised inland in the United States. Arctic char can also be raised onshore in systems that recirculate water, reducing disease transfer and pollution. All of these species are delicious alternatives to ocean-farmed species; most any fish—even salmon—can be farmed far away from sensitive marine habitats.

Even offshore aquaculture methods can be modified to reduce pollution. Mobile fish pens that move around over different areas are one new approach. Methods are being developed to recycle fish sewage, and new feed formulations are being developed that use smaller amounts of wild fish and replace it with vegetable protein—for example, from soy. Multitrophic aquaculture, or integrated farms, put salmon pens near farmed plants and animals that consume the salmon wastes, species that can later be marketed themselves. Seaweeds are efficient waste recyclers that can extract about 40% of the dissolved nutrients available. Seaweed can be grown on ropes dangling from rafts downstream from salmon pens and grow in the wastewater, primarily ammonia (excreted by salmon) and decaying food. Filter-feeding mussels can also be cultured nearby to grow on particles of excrement and food scraps.

What is "Green Chemistry?"

Plastics can be manufactured that are degradable. Microbes have been genetically engineered to produce biodegradable plastics, which could benefit the oceans. Standards for measuring how plastics break down in particular environments have emerged recently and are still in development. Comparisons among plastics are complicated by the fact that no one entity is recognized as setting those standards. Nevertheless, there is consensus on distinctions among the key terms "degradable," "biodegradable," and "compostable." Degradable means that chemical changes take place, maybe due to sunlight or heat, that alter a plastic's structure and properties, like clouding or fragmenting. Biodegradable more narrowly indicates that degradation results from naturally occurring microorganisms (bacteria, fungi, or algae), but makes no guarantee that the degradation products are nontoxic or make good compost. Compostable goes a step further: the microorganisms' breakdown products must yield CO_2, water, inorganic compounds, and biomass and leave no visible, distinguishable, or toxic residue. The only standard for biodegradation of plastics in the marine environment requires that within six months it must have disintegrated into pieces smaller than two millimeters and that biodegradation must have progressed so that 30% of the carbon has been converted into CO_2. Bioplastics can be manufactured from corn or sugarcane. Green chemistry may also be able to design less toxic or persistent compounds to replace ones currently in use.

Since climate change is such a major threat, are there any effective national and international policies to curb it?

It is recognized that climate change is the biggest threat to the world's oceans (to the land, wildlife, agriculture, and to human health) yet there have been no effective policies established largely because of expense. A possible way of dealing with it would be by establishing a carbon tax, but people in developed countries will likely have to modify their lifestyles. The countries that are the most responsible for carbon emissions (e.g. United States and Europe) over the years are not the same ones that are bearing the brunt of the effects (e.g. coastal low-lying countries like Bangladesh and small island nations). China has become a major emitter. Countries will have to agree on who is responsible for curtailing how much of their emissions. Meanwhile, global emissions during the first decade of this century grew nearly twice as fast as during the previous 30 years. Within the United States (apparently not so much elsewhere), powerful moneyed interests such as oil companies are promoting the idea that climate change is controversial within scientific circles. This is not the case—97% of climate scientists agree that it is happening and that we are responsible for it. The controversy exists in politics not science, but the media feel obligated to provide equal time for the deniers and skeptics as if they had as much credibility as the National Academy of Sciences, the American Association for the Advancement of Science and other esteemed scientific organizations. While President Obama has put forth a climate plan, the opposing party in US politics stands firmly against any policies to deal with this urgent issue.

While there has not been significant progress in mitigation, there are many efforts toward adaptation or resilience, especially after the devastation of Hurricane Sandy. Huge numbers of buildings and other infrastructure have been damaged or destroyed by powerful hurricanes and floods. Programs are being designed to reduce the vulnerability of coastal structures and water-resource infrastructure.

What steps can local and state governments take to reduce pollution?

Runoff

The federal government should, but has not led the fight against the massive pollution from runoff and other diffuse sources. Much of the degradation is from chemical fertilizers and animal feed lots. In the absence of federal leadership, state and local governments can take steps to reduce the amount of runoff. They can purchase property in vulnerable areas and turn it into open space (natural areas). They can develop programs to use nature's defenses (green infrastructure) to buffer and restore marshes and barrier islands that can absorb pollution and protect inland property. They can plant forest buffers, and plant street trees in urban settings to absorb stormwater and reduce the heat island effect. They can develop rain gardens and green roofs, and replace impervious paved areas with permeable surfaces that absorb rainfall. Stormwater management should include bioretention systems that capture stormwater and treat urban runoff. For effective stormwater management, they should use landscape vegetation and specially designed filters that remove bacteria, metals, nutrients and suspended solids naturally. They should eliminate illegal discharges that drain into stormwater systems, and initiate programs to pick up pet wastes. Municipalities with old sewage systems and combined sewer overflows (CSOs) should develop systems for storing water after heavy rainfalls. This can be done by creating bioswales, which are planted areas designed to collect and absorb stormwater. Each system (20 x 5 ft) can manage over 2,000 gallons of street and sidewalk runoff during a storm. Underground, they filter and store excess stormwater in layers of broken stone and soil. In 2012–2013, New York City constructed many in Queens, the Bronx, and Brooklyn, and more are being planned. This is an innovative way to address street flooding and reduce sewer overflows into nearby bodies of water such as Jamaica Bay, Newtown Creek, and the Gowanus Canal. Together, these projects will keep over seven million gallons of stormwater out of the sewer system annually. When combined sewer overflows do occur in Newtown Creek after rain, sensors can send twitter and text message alerts to residents of the watershed to limit their toilet flushing.

Green roofs are planted areas on roofs that serve to absorb rainwater, provide insulation, create a habitat for wildlife, and help to lower urban summer air temperatures to mitigate the heat island effect. There are two types of green roofs: intensive roofs, which are thicker and support a wider variety of plants but are heavy and require much maintenance; and extensive roofs, which are covered in a thinner layer of vegetation and are lighter. Switzerland, particularly Basel, has the highest area of green roofs per capita in the world. Since the 1990s, Switzerland has required every new building that has a suitable roof pitch to have a green roof—the building's owner must plant and maintain some kind of natural greenery. The use of green roofs was stimulated by financial incentives and building regulations. About 23% of Basel's flat roof area was green in 2006. For developers, installing green roofs is now routine, and developers do not object to installing them.

Toronto approved a law in May 2009 mandating green roofs on residential and industrial buildings.

Climate Change

For climate change, municipalities should utilize smart-growth principles to develop new neighborhoods that don't require automobiles and that maximize walking and bicycling and include as much open space as possible. They should put bike lanes along streets and make streets pedestrian- and bike-friendly. They should use and encourage citizens to use alternative fuels, improve public transportation with well-designed stations, and put new housing and businesses near train stations to spur use of public transportation. Municipal fleets should use alternative fuels (green fleets), and anti-idling policies should be established and enforced. High-efficiency, green-design commercial, residential, and municipal buildings should be constructed, preferably with green roofs. For buildings that cannot support green roofs, white roofs can reflect more sunlight and thus reduce warming, even if they don't have the additional advantages of green roofs. Solar panels on roofs of buildings in sunny areas should be encouraged. Wetlands should be built and restored that can absorb carbon as well as provide habitat and protect inland structures from storms. For coastal areas that are frequently flooded, managed retreat, or, as the UK's environmental agency refers to it, "managed coastal realignment" is a sensible response.

Emerging Concerns

For contaminants of emerging concern (CECs) sewage plants should be upgraded to advanced methods of sewage treatment. Sewage plants should require pretreatment or pollution prevention plans for facilities that are likely to release emerging contaminants, including hospitals, long-term care facilities, hospices, veterinary hospitals, and compounding pharmacies. Some states have legislation in the works to prohibit the manufacture, distribution, and sale of personal care cosmetic products containing plastic microbeads. State and national proposals to regulate toxic fire retardants should be supported and flammability standards should be changed so that toxic fire retardants such as PBDEs are not mandated in upholstery and furniture.

Debris

Effective waste reduction and recycling programs should be developed. In the Los Angeles area, 20 tons of plastic fragments—from grocery bags, straws, and soda bottles—are carried into the Pacific Ocean every day. The state is focusing on preventing garbage from entering the sewer systems in the cities in the watershed, because those sewers lead to the Los Angeles River. To keep garbage out of the sewers, the state is regularly cleaning the streets, educating the public about proper garbage disposal, and providing public garbage cans. In addition, the state is removing garbage that enters the sewers by using devices that sieve out much of the garbage from the sewers before it reaches the river. Getting the EPA and state agencies to set strict water-quality standards for plastic pollution will help promote early detection and prevention of plastic waste as well as the cleanup of beaches and oceans. It will encourage states and municipalities to develop new ways to limit the plastic entering the waste stream, and stimulate creative solutions to this pervasive problem.

Toxic Chemicals

In the absence of strong federal laws, state laws should be strengthened to reduce pollution from industrial chemicals. In California, Proposition 65 requires businesses to notify people about the chemicals that are in consumer products or are released into the environment. The list of chemicals includes many substances that are known to cause cancer, birth defects, or other reproductive harm, including additives or ingredients in pesticides, common household products, food, drugs, dyes, or solvents. By providing this information, Proposition 65 enables Californians to make informed decisions about their chemical exposure. It also prohibits businesses from knowingly discharging significant amounts of listed chemicals into sources of drinking water. This law has increased public awareness about adverse effects of chemical exposure and has provided an incentive for manufacturers to remove listed chemicals from their products. It has been responsible for reformulation of many consumer products to eliminate toxic chemicals.

In 2014 the Vermont Senate passed one of the toughest policies in the nation to regulate toxic chemicals found in all consumer products, but business lobbyists downsized the bill in the House, limiting the scope of the chemical reporting requirement to children's products. Currently, the state regulates chemicals one at time, and has done so only for mercury, bisphenol A (BPA), lead, and flame retardants.

Invasive Species

Monitoring should be done frequently and carefully, and when a known invasive species is detected forces should be mobilized quickly to eradicate it. In a classic example of better-late-than-never, the Florida Fish and Wildlife Commission in 2014 banned the import of lionfish for the aquarium trade. This action, taken many years after the invasion, was explained as a way to limit new introductions into Florida waters.

What actions can individual citizens take to reduce marine pollution?

Individual actions can make a big difference in reducing marine pollution.

Runoff

There are many things citizens can do to reduce nonpoint source pollution, including the following: for eutrophication and runoff, plant grass, trees and shrubs in bare areas. They will reduce and absorb runoff, and their roots will hold the soil together, reducing erosion and runoff. Use a rain barrel to capture some rainfall. Use fertilizers and pesticides sparingly. If you have property along the water, keep a natural shoreline with marsh vegetation that can absorb runoff, buffer the water from chemicals, and buffer the property from wind and storms. Volunteer for marsh restoration plantings or oyster gardening projects (oysters filter a lot of phytoplankton out of the water, reducing eutrophication). If you live in a tropical area, volunteer for mangrove or coral reef restoration projects if there are any. If you are a boater, use sewage pump-out facilities. Slow down in shallow areas to reduce boat wake erosion in areas with submerged aquatic vegetation, salt marshes, and wildlife. If you are a farmer, use conservation tillage or no-till, plant buffers at the edge of streams, use efficient irrigation systems, and manage manure. Grow lots of legumes (e.g., peas, beans, lentils), which use nitrogen from the air and don't need fertilizer. There are organic farms that use no commercial fertilizer, spray no chemicals, and reuse all their stormwater. A farmer named Walden (appropriate name) in York County, Pennsylvania uses animals to do the work. Chicken waste, from chickens in mobile pens on wheels, is used to fertilize the fields; pigs and cattle eat the weeds and graze on the grass.

Litter

Be careful to manage trash. Go on stream or beach walks, removing trash and debris. Recycle plastic, glass, and paper. Use alternatives to plastic bags. Go shopping with your own reusable bags. Avoid single-use, nonrecyclable plastics (cutlery, plates, disposable plastic or Styrofoam cups). Travel with a permanent coffee cup rather than constantly trashing plastic-lined paper or Styrofoam cups, and patronize businesses that avoid single-use plastics (e.g., go to a farmer's market rather than a store that gives away hundreds of plastic bags daily). Less trash generated means less ending up in the ocean. Keep trash out of storm drains, where it will either clog the drain or end up in the water.

Climate Change

Conserve energy in your home, and you will also save money and give less to the power company. Insulate your home so less heat escapes and you will save money. Use compact fluorescent light bulbs. Buy an energy-efficient automobile, but try to carpool or take public transportation whenever you can. Take a bicycle or walk as much as possible—it will also help you stay in shape. If your house is in a sunny area, consider putting solar panels on the roof or making a green roof. Reduce your carbon footprint by consuming less, and recycling and reusing more.

Invasive Species

If you tire of your fish tank or any resident in it, don't bring it to a nearby water body—take it to a pet store or find another home for it. If you have wading boots, wash the mud off before going to another location. If you fish, don't release any live bait organisms. If you are a gardener, focus on native plants.

Toxic Chemicals

Safe disposal of household hazardous wastes (e.g., oil, drugs, electronics, batteries) is important. Take unneeded paints, solvents, and pesticides to hazardous materials collection. Do not pour chemicals on the ground or into storm drains, where they will get into a stream or river. Don't throw out old batteries in the regular garbage. Take used motor oil to oil recycling facilities. To reduce pesticide use, control pests with beneficial insects such as ladybugs and praying mantises. Survey your yard to see what pests are there and then use pesticides only if natural predators cannot keep them in check.

Emerging Concerns

Do not flush unused pills down the toilet. Avoid products with CECs including paints, room fresheners, plastic shower curtains, clothes, and other items made with PVC fabric. Do not use antimicrobial hand cleaners containing triclosan or triclocarban. Warm water and soap or alcohol-based sanitizers are equally effective at removing germs. If you use exfoliants to massage your skin, check the ingredients. Also check the ingredients in your toothpaste, shampoo, and soap. If you see "polyethylene" or another type of plastic, you are releasing plastic microbeads into the water when you wash the product off. There are alternative products that use natural ingredients.

And in general, talk to people about ocean pollution. Consider becoming active by lobbying (e.g., writing letters to the editor, visiting politicians, advocating to chambers of commerce and rotaries for more recycling and strengthening policies to prevent water pollution) and joining groups whose policies you support.

What are the overall status and trends of marine pollution?

Overall, there is good news and bad news. Some of the major types of pollutants are decreasing, including persistent organic pollutants like DDT and PCBs, which are no longer being manufactured or used in most countries. Levels of trace metals, however, have not shown overall downward trends. The persistence of contaminants in bottom sediments is one reason that things are not improving as fast as we might like. New activities such as seabed mining, especially in vulnerable environments such as the deep ocean or the Arctic, are likely to produce new pollution unless preventative measures and regulatory frameworks are put in place beforehand. Getting out ahead of emerging threats before they become major problems would be a refreshing new approach to pollution.

Oil spills have been decreasing over the past two decades, and tankers built in the United States will have double hulls. In response to the Deepwater Horizon the government set up panels to provide expert advice to prevent future blowouts. A presidential commission recommended many measures to improve drilling safety. However, the government has not followed important recommendations for increasing safety, such as the design of the blowout preventer which could not stop blowouts in deep water. There are plans to expand offshore drilling into deeper waters, making a future disaster more likely.

Other types of pollution have been increasing, including nutrients that cause eutrophication, as indicated by the increasing number of hypoxic zones and harmful algal blooms around the world. In the future, there may be greater controls on point sources that will result in further decreases of persistent organic pollutants and sewage wastes. Whether regulations will be developed to reduce nonpoint sources of pollution is an open question. More recently recognized pollutants, including flame retardants, pharmaceuticals, and nanoparticles, will become increasingly important sources of toxicity until they are eventually—hopefully sooner rather than later—controlled by regulations. Marine debris is an increasing problem that has not yet been controlled. Invasive species will probably continue to arrive in new locations, although ballast water regulations will reduce the importance of this vector. Climate change will exacerbate effects of toxic contaminants; warmer water will increase metabolic rates of marine organisms, which will generally increase toxic effects. Increased temperature will cause huge changes in the ocean ecosystems, and the lowered pH from ocean acidification will cause its own harmful effects, especially on shell-forming species. Climate change will likely also exacerbate the effects of eutrophication, as warmer water will hold less oxygen and remain stratified longer, intensifying hypoxia in deeper waters. Warmer water is likely to be conducive to blooms of nuisance algae.

Healthy oceans provide food, jobs, and recreation for large numbers of people and are a potential source of clean energy and new medications. There is no shortage of international recommendations and action plans for restoring the oceans' health. Countless reports have come out from different organizations that sound an alarm about the state of the oceans and call for action—with little response from the powers that be. Individual countries cannot do it alone but have to cooperate since pollution does not honor national boundaries, nor does marine life. A 2013 report from the International Programme on the State of the Ocean (IPSO), a nongovernmental group of leading scientists, has concluded that the world's oceans are under greater threat than previously believed from a deadly trio of global warming, declining oxygen levels, and acidification. The report indicated that conditions are ripe for the sort of mass extinction event that has happened in the past, but for the most part, the public and policymakers are failing to recognize—or are ignoring—the severity of the situation. The report makes it clear that deferring action will increase costs in the future and lead to even greater, perhaps irreversible, losses. These findings, while not really new, are a cause for alarm, as well as a blueprint for action. In order to protect the worlds' oceans and their resources that we depend on, it is vital that nations and the international community take major steps to reduce inputs of marine pollutants and reduce greenhouse gases. In 2014, the Intergovernmental Panel on Climate Change reported that global emissions of CO_2 rose 2.2% annually between 2000 and 2010.

To paraphrase Bob Dylan's famous civil rights and anti-war song, "How many floods will it take...? and how many dead coral reefs will it take...?" Let us hope the answer is rapid and effective and not just "blowin' in the wind."

REFERENCES

Chapter 1

Carson, R. 1951. The Sea Around Us. Oxford University Press, 278 pp.

Halpern, B. S. et al. 2008. A global map of human impact on marine ecosystems. *Science* 319: 948–952.

International Maritime Organization (IMO) International Convention for the Prevention of Pollution from Ships (MARPOL). www.imo. org/About/Conventions/ListOfConventions/Pages/International-Convention-for-the-Prevention-of-Pollution-from-Ships-%28MARPOL%29.aspx

Long, E. R. 2000. Degraded sediment quality in US estuaries: A review of magnitude and ecological implications. *Ecological Applications* 10: 338–349.

Rowe, C. L. 2008. The calamity of so long life: life histories, contaminants, and potential emerging threats to long-lived vertebrates. *BioScience* 58: 623–631.

US EPA Summary of the Clean Water Act. www2.epa.gov/laws-regulations/summary-clean-water-act

Chapter 2

Anderson, D. M. 2004. The growing problem of harmful algae. *Oceanus Magazine*, Woods Hole Oceanographic Institute, Woods Hole, MA. www.whoi.edu/page.do?pid=11913&tid=282&cid=2483.

Anderson, D. M., B. Reguera, G. C. Pitcher, and H. O. Enevoldsen 2010. The IOC international harmful algal bloom program: History and science of impacts. *Oceanography* 23: 72–85.

Boesch, D. F. et al. 1997. Harmful algal blooms in coastal waters: Options for prevention, control and mitigation. NOAA Coastal Ocean Program, Decision Analysis Series No. 10, Silver Spring, MD.

Deegan, L. A. et al. 2012. Coastal eutrophication as a driver of salt marsh loss. *Nature* 490: 388–392.

Diaz, R. J. and R. Rosenberg 2008. Spreading dead zones and consequences for marine ecosystems. *Science* 321:926–929.

Diaz, R. J., M. Selman, and C. Chique-Canache 2010. Global eutrophic and hypoxic coastal systems: Eutrophication and hypoxia—nutrient pollution in coastal waters. World Resources Institute, Washington, DC. www.wri.org/project/eutrophication

Rabalais, N. N. et al. 2010. Dynamics and distribution of natural and human-caused hypoxia. *Biogeosciences* 7: 585–619.

Selman, M., Z. Sugg, S. Greenhalgh, and R. Diaz 2008. Eutrophication and hypoxia in coastal areas: A global assessment of the state of knowledge. World Resources Institute, Washington DC. www.wri.org/publication/eutrophication-and-hypoxia-in-coastal-areas.

Chapter 3

Avery-Gomm, S. et al. 2012 Northern fulmars as biological monitors of trends of plastic pollution in the eastern North Pacific. *Marine Pollution Bulletin* 64: 1776–1781.

Bergmann, M. and M. Klages 2012. Increase of litter at the Arctic deep-sea observatory HAUSGARTEN. *Marine Pollution Bulletin* 64: 2734–2741.

Bilkovic, D. M., K. J. Havens, D. M. Stanhope, and K. T. Angstadt 2012. Use of fully biodegradable panels to reduce derelict pot threats to marine fauna. *Conservation Biology* 26: 957–966.

Browne, M. A., S. J. Niven, T. S. Galloway, S. J. Rowland, and R. C. Thompson 2013. Microplastic moves pollutants and additives to worms, reducing functions linked to health and biodiversity. *Current Biology* 23 (23): 2388 DOI: 10.1016/j.cub.2013.10.012.

Galgani, F., S. Jaunet, A. Campillo, X. Guenegen, and E. His. 1995. Distribution and abundance of debris on the continental shelf of the north-western Mediterranean Sea. *Marine Pollution Bulletin* 30: 713–717.

Hammer, J., M. H. Kraak, and J. R. Parsons 2012. Plastics in the marine environment: The dark side of a modern gift. *Reviews of Environmental Contamination and Toxicology* 220: 1–44.

Laist, D. W. 1997. Impacts of marine debris: Entanglement of marine life in marine debris including a comprehensive list of species with entanglement and ingestion records. *In*: Marine debris: Sources, impacts, and solutions. Ed. J. M. Coe and D. B. Rogers. Springer Verlag, NY, pp 99–140.

Law, K. L. et al. 2010. Plastic accumulation in the North Atlantic subtropical gyre. *Science* 329: 1185–1188.

Martinez, E., K. Maamaatuaiahutapu, and V. Taillandier 2009. Floating marine debris surface drift: Convergence and accumulation toward the South Pacific subtropical gyre. *Marine Pollution Bulletin* 58: 1347–1355.

Moore Recycling Associates 2012. Plastic Recycling Collection: National Research Study: 2012 Update. http://www.moorerecycling.com/ UpdatedREACHReportMay2013.pdf

National Academies of Science 2008. Tackling marine debris in the 21st century. National Academies Press, Washington, DC. 218 pp.

Ocean Conservancy 2013. Ocean Trash Index. http://www.oceanconservancy.org/our-work/international-coastal-cleanup/2012-ocean-trash-index.html.

Schlining, K. S. et al. 2013. Debris in the deep: Using a 22-year video annotation database to survey marine litter in Monterey Canyon, central California, USA. Deep Sea Research Part I: Oceanographic Research Papers, 2013; DOI: 10.1016/j.dsr.2013.05.006.

United Nations Environment Programme (UNEP) 2011. UNEP Year Book 2011, Plastic Debris in the Ocean. www.unep.org/yearbook/2011/pdfs/plastic_debris_in_the_ocean.pdf.

US Environmental Protection Agency (EPA) and National Oceanic and Atmospheric Administration (NOAA); J.-P. Gattuso (Topic Editor) 2007. Marine Debris. In: Encyclopedia of Earth. Ed. C. Cleveland. Washington DC, National Council for Science and the Environment.

Wright, S. L., D. Rowe, R. C. Thompson, and T. S. Galloway 2013. Microplastic ingestion decreases energy reserves in marine worms. *Current Biology* 23(23): R1031 DOI: 10.1016/j.cub.2013.10.068.

Chapter 4

Culbertson, J. B., I. Valiela, E. E. Peacock, C. M. Reddy, A. Carter, and R. VanderKruik 2007. Long-term biological effects of petroleum residues on fiddler crabs in salt marshes. *Marine Pollution Bulletin* 54: 955–962.

Dissanayake, A., C. Piggott, C. Baldwin, and K. A. Sloman 2010. Elucidating cellular and behavioural effects of contaminant impact (polycyclic aromatic hydrocarbons, PAHs) in both laboratory-exposed and field-collected shore crabs, *Carcinus maenas* (Crustacea: Decapoda). *Marine Environmental Research* 70: 368–373.

Dubansky, B., A. Whitehead, J. Miller, C. D. Rice, and F. Galvez 2013. Multi-tissue molecular, genomic, and developmental effects of the Deepwater Horizon oil spill on resident Gulf killifish (*Fundulus grandis*). *Environmental Science & Technology* 47: 5074–5082.

Etkin, D. S. 2001. Analysis of oil spill trends in the United States and worldwide. Paper presented at 2001 International Oil Spill Conference, pp 1292–1300.

Heintz, R., J. W. Short, and S. D. Rice 1999. Sensitivity of fish embryos to weathered crude oil: Part II. Increased mortality of pink salmon (*Oncorhynchus gorbuscha*) embryos incubating downstream from weathered *Exxon valdez* crude oil. *Environmental Toxicology and Chemistry* 18: 494–503.

Huijer, K. 2005. Trends in oil spills from tanker ships, 1995–2004. Paper presented at the 28th Arctic and Marine Oilspill Program (AMOP) Technical Seminar, Calgary, Canada.

Incardona, J., et al. 2012. Unexpectedly high mortality in Pacific herring embryos exposed to the 2007 *Cosco Busan* oil spill in San Francisco Bay. *Proceedings of the National Academy of Sciences* 109 (2): E51–E58

Incardona, J., et al. 2014. *Deepwater Horizon* crude oil impacts the developing hearts of large predatory pelagic fish. *Proceedings of the National Academy of Sciences* 111: E1510–E1518. doi:10.1073/pnas.1320950111

Myers, M. S., J. T. Landahl, M. M. Krahn, L. L. Johnson, and B. B. McCain 1990. Overview of studies on liver carcinogenesis in English sole from Puget Sound; evidence for a xenobiotic chemical etiology I: Pathology and epizootiology. *Science of the Total Environment* 94: 33–50

Patin, S. 2008. Oil spill. *In:* Encyclopedia of Earth. Ed. C. J. Cleveland. Washington, DC: Environmental Information Coalition, National Council for Science and the Environment. www.eoearth.org/article/Oil_spill.

Peterson, C. H., S. D. Rice, J. W. Short, D. Esler, J. L. Bodkin, B. E. Ballachey, and D. B. Irons 2003. Long-term ecosystem response to the *Exxon Valdez* oil spill. *Science* 302: 2082–2086.

Sanders, H. L., J. F. Grassle, G. R. Hampson, L. S. Morse, S. Garner-Price, and C. C. Jones 1980. Anatomy of an oil spill: Long-term effects from the grounding of the barge *Florida* off West Falmouth, Massachusetts. *Journal of Marine Research* 38: 265–324.

Chapter 5

Alzieu, C. 1990. Environmental impact of TBT: The French experience. *Science of the Total Environment* 258: 99–102.

Boucher, O. et al. 2012. Prenatal methylmercury, postnatal lead exposure, and evidence of attention deficit/hyperactivity disorder among Inuit children in Arctic Québec. *Environmental Health Perspectives* 120: 1456–1461.

Gibbs, P. E. and G. W. Bryan 1986. Reproductive failure in populations of the dog-whelk, *Nucella lapillus*, caused by imposex induced by tributyltin from antifouling paints. *Journal of the Marine Biological Association of the United Kingdom* 66: 767–777.

Hansen, J. A., J. D. Rose, R. A. Jenkins, K. G. Gerow, and H. L. Bergman 1999. Chinook salmon (*Oncorhynchus tshawytscha*) and rainbow trout (*Oncorhynchus mykiss*) exposed to copper: Neurophysiological and histological effects on the olfactory system. *Environmental Toxicology and Chemistry* 18: 1979–1991.

Kirk, J. L. et al. 2012. Mercury in Arctic marine ecosystems: Sources, pathways, and exposure. *Environmental Research* 119: 64–87.

Klerks, P. and J. S. Weis 1987. Genetic adaptation to heavy metals in aquatic organisms: A review. *Environmental Pollution* 45: 173–206.

Kroglund, F. and B. Finnstad 2003. Low concentrations of inorganic monomeric aluminum impair physiological status and marine survival of Atlantic salmon. *Aquaculture* 222: 119–133.

Levinton, J. S., E. Suatoni, W. Wallace, R. Junkins, B. Kelaher, and B. J. Allen 2003. Rapid loss of genetically based resistance to metals after the cleanup of a Superfund site. *Proceedings of the National Academy of Sciences* 100: 9889–9891.

Polak-Juszczak, L. 2009. Temporal trends in the bioaccumulation of trace metals in herring, sprat, and cod from the southern Baltic Sea in the 1994–2003 period. *Chemosphere* 76: 1334–1339.

Weis, J. S., G. Smith, T. Zhou, C. Bass, and P. Weis 2001. Effects of contaminants on behavior: biochemical mechanisms and ecological consequences. *BioScience* 51: 209–218.

Weis, J. S. and P. Weis 2004. Metal uptake, transport, and release by wetland plants: Implications for phytoremediation and restoration. *Environment International* 30: 685–700.

Chapter 6

Carson, R. 1962. *Silent Spring*. Houghton Mifflin, NY. 381 pp.

Colborn T., F. S. vom Saal, and A. M. Soto 1993. Developmental effects of endocrine-disrupting chemicals in wildlife and humans. *Environmental Health Perspectives* 101: 378–384.

Chapman, P. 1990. The sediment quality triad approach to determining pollution-induced degradation. *Science of the Total Environment* 97–98: 815–825.

De Guise, A., A. Lagace, and P. Beland 1994. True hermaphroditism in a St. Lawrence beluga whale (*Delphinapterus leucas*). *Journal of Wildlife Diseases* 30: 287–290.

Feldman, K., D. Armstrong, B. R. Dumbould, T. H. DeWitt, and D. Doty 2000. Oysters, crabs and burrowing shrimp: Review of an aquatic conflict over aquatic resources and pesticide use in Washington State's (USA) coastal estuaries. *Estuaries* 23:141–176.

Hudson River Natural Resource Trustees 2013. PCB Contamination of the Hudson River ecosystem: Compilation of contamination data through 2008. 38 pp. www.fws.gov/contaminants/restorationplans/HudsonRiver/docs/Hudson%20River%20Status%20Report%20Update%20January%202013.pdf.

King, R. S., J. R. Beaman, D. F. Whigham, A. H. Hines, M. E. Baker, and D. E. Weller 2004. Watershed land use is strongly linked to PCBs in white perch in Chesapeake Bay subestuaries. *Environmental Science and Technology* 38: 6546–6552.

Lykousis, V. and M. Collins 2005. Impact of natural and trawling events on resuspension, dispersion and fate of pollutants. (INTERPOL). Introduction to Special Issue. *Continental Shelf Research* 25: 2309–2314.

Matthiessen, P. and P.E. Gibbs 1998. Critical appraisal of the evidence for tributyltin-mediated endocrine disruption in mollusks. *Environmental Toxicology and Chemistry* 17: 37–43.

Nacci, D. et al. 1999. Adaptations of wild populations of the estuarine fish *Fundulus heteroclitus* to persistent environmental contaminants. *Marine Biology* 134: 9–17.

NOAA Center for Coastal Monitoring and Assessment. Mussel Watch Contaminant Monitoring. http://ccma.nos.noaa.gov/about/coast/nsandt/musselwatch.aspx.

Scholz, N. L. et al. 2012. A perspective of modern pesticides, pelagic fish declines, and unknown ecological resilience in highly managed ecosystems. *BioScience* 62: 428–434.

Chapter 7

André M. et al. 2011. Low-frequency sounds induce acoustic trauma in cephalopods. *Frontiers in Ecology and the Environment* 9: 489–493.

Baun, A., N. B. Hartman, K. Grieger, and K. O. Kusk 2008. Ecotoxicity of engineered nanoparticles to aquatic invertebrates—A brief review and recommendations for future toxicity testing. *Ecotoxicology* 17: 387–395.

Bay, S. et al. 2011. Sources and effects of endocrine disruptors and other contaminants of emerging concern in the Southern California Bight coastal ecosystem. Technical Report 650 from the Southern California Coastal Water Research Project (SCCWRP), pp 11–20.

Beulig, A., and J. Fowler 2008. Fish on prozac: Effect of serotonin reuptake inhibitors on cognition in goldfish. *Behavioral Neuroscience* 122: 426–432.

Cleveland, D. et al. 2012. Pilot estuarine mesocosm study on the environmental fate of Silver nanomaterials leached from consumer products. *Science of the Total Environment* 421–422: 267–272.

Daughton, C.D. and T.A. Ternes 1999. Pharmaceuticals and personal care products in the environment: agents of subtle change? *Environmental Health Perspectives* 107 Supplement 6: 907–938.

Jobling, S., M. Nolan, D. R. Tyler, F. Brightly, and J. P. Sumpter 1998. Widespread sexual disruption in wild fish. *Environmental Science and Technology* 32: 2498–2506.

Kelly, B. et al. 2009. Perfluoroalkyl contaminants in an Arctic marine food web: Trophic magnification and wildlife exposure. *Environmental Science and Technology* 43: 4037–4043.

Madigan, D. J., Z. Baumann, and N. S. Fisher 2012. Pacific bluefin tuna transport Fukushima-derived radionuclides from Japan to California. *Proceedings of the National Academy of Sciences* 109: 9483–9486

Purdom, C. E., P. A. Hardiman, V. J. Bye, N. C. Eno, C. R. Tyler, and J. P. Sumpter 1994. Estrogenic effects of effluents from sewage treatment works. *Journal of Chemical Ecology* 8: 275–285.

Reiner, J. S., G. O'Connell, A. J. Moors, J. R. Kucklick, P. R. Becker, and J. M. Keller 2011. Spatial and temporal trends of perfluorinated compounds in Beluga whales (*Delphinapterus leucas*) from Alaska. *Environmental Science and Technology* 45: 8129–8136.

US EPA Pharmaceuticals and Personal Care Products (PPCPs). Frequently Asked Questions. http://epa.gov/ppcp/faq.html_www.nhs.uk/news/2008/11November/Pages/NanoparticleQA.aspx.

Weingart, L. S. 2007. The impacts of anthropogenic ocean noise on cetaceans and implications for management. *Canadian Journal of Zoology* 85: 1091–1116.

Chapter 8

Bergey, L. and J.S. Weis 2007. Molting as a mechanism of depuration of metals in the fiddler crab, *Uca pugnax. Marine Environmental Research* 64: 556–562.

Bignert, A. 2004. PCB concentration in fish muscle. Helsinki Convention. www.helcom.fi/environment2/ifs/archive/ifs2004/en_GB/pcbfish/.

Bignert, A., E. Nyberg, and S. Danielsson 2007. Lead concentrations in fish liver. Helsinki Convention. www.helcom.fi/environment2/ifs/ifs2007/en_GB/lead_fish/.

Elliott, J. E., and K. H. Elliott 2012. Tracking marine pollution. *Science* 340: 556–558.

Fein, G. G., J. L. Jacobson, S. W. Jacobson, P. M. Schwartz, and Jeffrey K. Dowler 1984. Prenatal exposure to polychlorinated biphenyls: Effects on birth size and gestational age. *Journal of Pediatrics* 105: 315–320.

Ministry of Environment of British Columbia 2007. Environmental trends: Long-term trends in persistent organic pollutants in bird eggs. www.env.gov.bc.ca/soe/et07/05_contaminants/technical_paper/contaminants.pdf.

National Oceanic and Atmospheric Administration (NOAA) 2008. National status and trends mussel watch program: An assessment of two decades of contaminant monitoring in the nation's coastal zone from 1986–2005. Washington, DC.

Pohl, C. and U. Hennings 2008. Trace metals in Baltic seawater. *In*: State and Evolution of the Baltic Sea 1952-2005. Ed. R. Feistel, G. Nausch, and N. Wasmund. John Wiley & Sons, NJ, pp 367–393.

Rotkin-Ellman, M., K. K. Wong, and G. M. Solomon 2012. Seafood contamination after the BP Gulf oil spill and risks to vulnerable populations: A critique of the FDA risk assessment. *Environmental Health Perspectives* 120: 157–161.

Von Westernhagen, H., P. Cameron, D. Janssen, and M. Kerstan 1995. Age and size dependent chlorinated hydrocarbon concentrations in marine teleosts. *Marine Pollution Bulletin* 30: 655–659.

Wallace, W. G. and S. N. Luoma 2003. Subcellular compartmentalization of Cd and Zn in two bivalves. II. Significance of trophically available metal (TAM). *Marine Ecology Progress Series* 257: 125–137.

Chapter 9

Baumann, H., S. C. Talmage, C. J. Gobler, 2012. Reduced early life growth and survival in a fish in direct response to increased carbon dioxide. *Nature Climate Change* 2: 38–41.

Cazenave, A. and W. Llovel, 2010. Contemporary sea level rise. *Annual Review of Marine Science* 2: 145–173.

Cooley, S. R., N. Lucey, H. Kite-Powell, and S. C. Doney 2012. Nutrition and income from molluscs today imply vulnerability to ocean acidification tomorrow. *Fish and Fisheries* 13: 182–215.

De'ath, G., K. E. Fabricius, H. Sweatman, and M. Puotinen 2012. The 27-year decline of coral cover on the Great Barrier Reef and its causes. *Proceedings of the National Academy of Sciences* 44: 17995–17999.

De la Haye, K. L., J. Spicer, S. Widdicombe, and M. Briffa 2012. Reduced pH sea water disrupts chemo-responsive behaviour in an intertidal crustacean. *Journal of Experimental Marine Biology and Ecology* 412: 134–140.

Dyurgerov, M.B., and M. F. Meier 2005. Glaciers and the changing earth system: A 2004 snapshot. Occasional Paper 58. Institute of Arctic and Alpine Research, University of Colorado, Boulder, CO.

Form, A. U. and U. Riebesell 2012. Acclimation to ocean acidification during long-term CO_2 exposure in the cold-water coral *Lophelia pertusa*.*Global Change Biology* 18: 843–853.

Forster, J., A. G. Hirst, and D. Atkinson. 2012. Warming-induced reductions in body size are greater in aquatic than terrestrial species. *Proceedings of the National Academy of Sciences* 109: 19310–19314.

Hoegh-Guldberg, O. and J. F. Bruno 2010. The impact of climate change on the world's marine ecosystems. *Science* 328: 1523–1528.

Hoegh-Guldberg, O. et al. 2007. Coral reefs under rapid climate change and ocean acidification. *Science* 318: 1737–1742.

Intergovernmental Panel on Climate Change (IPCC) 2011. Summary for policymakers. *In*: Intergovernmental panel on climate change special report on managing the risks of extreme events and disasters to advance climate change adaptation Eds. Field, C. B. et al. Cambridge University Press, Cambridge.

International Programme on the State of the Ocean 2013. OPSO State of the Ocean Report. www.stateoftheocean.org/index.cfm

Munday, P. L., A. J. Cheal, D. L. Dixson, J. L. Rummer, and K. E. Fabricius 2014. Behavioural impairment in reef fishes caused by ocean acidification at CO2 seeps. Nature Climate Change DOI: 10.1038/NCLIMATE2195.

Narita, D., K. Rehdanz, and R. Tol 2012. Economic costs of ocean acidification: a look into the impacts on global shellfish production. *Climatic Change* 113: 1049–1063.

Parker, L. M., P. M. Ross, W. A. O'Connor, L. Borysko, D. A. Raftos, and H. O. Pörtner 2012. Adult exposure influences offspring response to ocean acidification in oysters. *Global Change Biology* 18: 82–92.

Rignot, E. 2008. Changes in West Antarctic ice dynamics observed with ALOS PALSAR. *Geophysical Research Letters* 35, L12505. DOI: 10.1029/2008GL033365.

Schofield, O., H. W. Ducklow, D. G. Martinson, M. P. Meredith, M. A. Moline, and W. R. Fraser 2010. How do polar marine ecosystems respond to rapid climate change? *Science* 328:1520–1523.

Sunda, W. G. and W-J Cai 2012. Eutrophication induced CO2-acidification of subsurface coastal waters: Interactive effects of temperature, salinity, and atmospheric pCO_2. *Environmental Science & Technology* 46: 10651–10659.

Taylor, G. T. et al. 2012. Ecosystem responses in the southern Caribbean Sea to global climate change. *Proceedings of the National Academy of Sciences* 109: 19315–19320.

Chapter 10

Albins, M. 2013. Effects of invasive Pacific red lionfish *Pterois volitans* versus a native predator on Bahamian coral-reef fish communities. *Biological Invasions* 5: 29–43.

Bertness, M. D. and T. C. Coverdale 2013. An invasive species facilitates the recovery of salt marsh ecosystems on Cape Cod. Ecology. http://dx.doi.org/10.1890/12-2150.1.

Clark P. F., P. Campbell, B. Smith, P. S. Rainbow, D. Pearce, and R. P. Miguez 2008. The commercial exploitation of Thames mitten crabs: a feasibility study. A report for the Department for Environment, Food and Rural Affairs by the Department of Zoology, the Natural History Museum, London. DEFRA reference FGE 274. pp. 1–81 + appendices 1–6.

Carlton, J. T. and J. B. Geller 1993. Ecological roulette: The global transport of marine organisms. *Science* 261: 78–82.

Cohen, A. and J. T. Carlton 1998. Accelerating invasion rate in a highly invaded estuary. *Science* 279: 555–558.

Daehler, C. and D. Strong 1996. Status, prediction and prevention of introduced cordgrass *Spartina* spp. invasions in Pacific estuaries, USA. *Biological Conservation* 78: 51–58

Donovan, E. P. et al. 2007. Risk of gastrointestinal disease associated with exposure to pathogens in the sediments of the Lower Passaic River. *Applied and Environmental Microbiology* 74: 1004–1018.

Frazer, T. K., C. A. Jacoby, M. A. Edwards, S. C. Barry, and C. M. Manfrino 2012. Coping with the lionfish invasion: Can targeted removals yield beneficial effects? *Reviews in Fisheries Science* 20: 185–191.

Green, S. J., J. L. Akins, A. Maljković, and I. M. Côté 2012. Invasive lionfish drive Atlantic coral reef fish declines. *PLoS ONE* 7(3): e32596. www.plosone.org/article/info%3Adoi%2F10.1371%2Fjournal. pone.0032596.

Grosholz, E. and G. Ruiz 1995. Spread and potential impact of recently introduced European green crab, *Carcinus maenas*, in Central California. *Marine Biology* 122: 239–247.

Johnston, C. A. and R. N. Lipcius 2012. Exotic macroalga *Gracilaria vermiculophylla* provides superior nursery habitat for native blue crab in Chesapeake Bay. *Marine Ecology Progress Series* 467:137–146.

Jousson, O., J. Pawlowsky, L. Zaninetti, A. Meinesz, and C. F. Boudouresque 1998. Molecular evidence for the aquarium origin of the green alga *Caulerpa taxifolium* introduced to the Mediterranean Sea. *Marine Ecology Progress Series* 172: 275–280.

Lambert, G. 2009. Adventures of a sea squirt sleuth: Unraveling the identity of *Didemnum vexillum* a global ascidian invader. *Aquatic Invasions* 4: 5–28.

McDermott, J. J. 1991. A breeding population of the Western Pacific crab *Hemigrapsus sanguineus* (Crustacea: Decapoda: Grapsidae) established on the Atlantic Coast of North America. *Biological Bulletin* 181: 195–198.

Miller, M. A. et al. 2002. Coastal freshwater runoff is a risk factor for *Toxoplasma gondii* infection of southern sea otters (*Enhydra lutris nereis*). *International Journal for Parasitology* 32: 997–1006

North Carolina Sea Grant 2013. Ingesting invaders: Serving up lionfish. *Coastwatch* 4: 35–37.

Nuñez, M. A., S. Kuebbing, R. D. Dimarco, and D. Simberloff 2012. Invasive species: to eat or not to eat, that is the question. *Conservation Letters* 5: 334–341.

Ruiz, G., J. Carlton, E. Grosholz, and A. Hines 1997. Global invasions of marine and estuarine habitats by non-indigenous species: Mechanisms, extent, and consequences. *Integrative and Comparative Biology* 37: 621–632.

Shiganova, T. 1998. Invasion of the Black Sea by the ctenophore *Mnemiopsis leidyi* and recent changes in pelagic community structure. *Fisheries Oceanography* 7: 305–310.

Chapter 11

Bosch, D. F., R. H. Burroughs, J. E. Baker, R. P. Mason, C. L. Rowe, and R. L. Siefert 2001. Marine pollution in the United States. Prepared for the Pew Oceans Commission, 55 pp. www.pewtrusts.org/. . . / Reports/. . . ocean. . . /env_pew_oceans_pollution.pdf.

Copeland, C. 2010. Ocean dumping act: A summary of the law. Congressional Research Service, CRS Report to Congress, 9 pp www.gc.noaa.gov/documents/gcil_crs_oda.pdf.

Dylan, Bob 1963. "Blowin' in the Wind" www.bobdylan.com/us/node/25835

Halpern, B. et al. 2012. An index to assess the health and benefits of the global ocean. *Nature*. DOI: 10.1038/nature11397 www.oceanhealth-index.org/Countries.

International Programme on the State of the Oceans 2013. The state of the ocean 2013: Perils, prognoses and proposals. www.stateofthe-ocean.org/research.cfm.

NOAA National Status and Trends Program. Mussel watch program. NOAA National Status and Trends. An assessment of two decades of contaminant monitoring in the Nation's Coastal Zone. http://ccma.nos.noaa.gov/publications/MWTwoDecades.pdf.

US Environmental Protection Agency. A summary of the Clean Water Act. www2.epa.gov/laws-regulations/summary-clean-water-act.

US Environmental Protection Agency. Comprehensive Environmental Response, Compensation, and Liability Act (CERCLA) Overview. www.epa.gov/superfund/policy/cercla.htm.

US Environmental Protection Agency. Clean Water Act Section 319. http://water.epa.gov/polwaste/nps/cwact.cfm.

US Environmental Protection Agency. National Pollution Discharge Elimination System (NPDES). http://cfpub.epa.gov/npdes/cwa.cfm?program_id=45.

US Environmental Protection Agency. Summary of the Marine Protection, Research and Sanctuaries Act. www2.epa.gov/laws-regulations/summary-marine-protection-research-and-sanctuaries-act.

INDEX

Page numbers followed by *f* indicate figures.